DATE DUE

AC 9 02			
MR 0 02			

DEMCO 38-296

Educating for OSHA
Savvy Chemists

ACS SYMPOSIUM SERIES **700**

Educating for OSHA Savvy Chemists

Paul J. Utterback, EDITOR
Oregon Institute of Technology

David A. Nelson, EDITOR
University of Wyoming

Developed from a symposium sponsored by the Divisions
of Chemical Health and Safety and Chemical Education
at the 213th National Meeting
of the American Chemical Society,
San Francisco, California,
April 13–17, 1997

American Chemical Society, Washington, DC

Library of Congress Cataloging-in-Publication Data

Educating for OSHA Savvy Chemists / Paul J. Utterback, editor, David A.
   Nelson, editor.

   p.   cm.—(ACS symposium series, ISSN 0097–6156; 700)

   "Developed from a symposium sponsored by the Division of Chemical
Health and Safety and Chemical Education at the 213th National Meeting of the
American Chemical Society, San Francisco, California, April 13–17, 1997."

   Includes bibliographical references and indexes.

   ISBN 0–8412–3569–4

   1. Chemical laboratories—Safety measures—Study and teaching (Higher)—
United States—Congresses.

   I. Utterback, Paul J. II. Nelson, David A. (David Alan), 1931–   . III.
American Chemical Society. Division of Chemical Health and Safety. IV.
American Chemical Society. Division of Chemical Education. V. American
Chemical Society. Meeting (213th : 1997 : San Francisco, Calif.) VI. Series.

QD63.5.E38   1998
542′.1′0289—dc21

                                                                98–16896
                                                                      CIP

This book is printed on acid-free, recycled paper.

Foreword

THE ACS SYMPOSIUM SERIES was first published in 1974 to provide a mechanism for publishing symposia quickly in book form. The purpose of the series is to publish timely, comprehensive books developed from ACS sponsored symposia based on current scientific research. Occasionally, books are developed from symposia sponsored by other organizations when the topic is of keen interest to the chemistry audience.

Before agreeing to publish a book, the proposed table of contents is reviewed for appropriate and comprehensive coverage and for interest to the audience. Some papers may be excluded in order to better focus the book; others may be added to provide comprehensiveness. When appropriate, overview or introductory chapters are added. Drafts of chapters are peer-reviewed prior to final acceptance or rejection, and manuscripts are prepared in camera-ready format.

As a rule, only original research papers and original review papers are included in the volumes. Verbatim reproductions of previously published papers are not accepted.

ACS BOOKS DEPARTMENT

Contents

TECHNIQUES FOR ACHIEVING AN
OSHA SAVVY CHEMIST

INDEXES

Preface

Colleges and Universities, as well as chemical industries, face increasing expectations by both the public and regulatory agencies to perform in accordance with accepted chemical safety standards. Graduating chemists should keep pace with this scrutiny by learning safe handling and disposal of chemicals. This educational initiative is pursued by many schools, colleges, and universities by upgrading chemical safety education within the chemistry curriculum.

The Spring 1997 American Chemical Society (ACS) Symposium *"Educating for OSHA Savvy Chemists"* resulted from a history of cooperation between the divisions of Chemical Health and Safety and Chemical Education. Previous Symposia focused the need for chemical safety education. The Spring 1997 Symposium brought three important groups together to present ideas on why OSHA Savvy was needed, how to achieve departmental-level compliance with chemical regulations, and specific educational techniques to include safety principles within the chemistry curriculum.

Because the Symposium represented a conglomeration of new ideas and advice from a variety of sources it became clear that a published volume would benefit a varied audience. Whether written by a new student, practicing chemist, or chemical educator the chapters presented here provide invaluable resources for addressing the evolving world of safe chemical practice.

A major objective of this symposium volume is to serve as a guide for designing plans to provide chemical professionals with the basic education to understand the relationships between the hazardous properties of the chemicals with which they work and chemical health and safety issues, government regulations, and industrial hygiene in the chemical field. It does not cover all related issues, specify all minimum legal standards, or necessarily represent the policy of the ACS.

PAUL J. UTTERBACK
Tacoma Public Utilities
P.O. Box 11007
Tacoma, WA 98411

DAVID A. NELSON
Department of Chemistry
University of Wyoming
Box 3838, University Station
Laramie, WY 82071

Why Educate for OSHA Savvy Chemists?

Chapter 1

Educating for OSHA Savvy Chemists: The Role of Academia in a Shared Responsibility

Paul J. Utterback[1] and David A. Nelson[2]

[1]Department of Safety and Environmental Health, Oregon Institute of Technology, Klamath Falls, OR 97601
[2]Department of Chemistry, University of Wyoming, Laramie, WY 82071

Understanding the scientific basis of chemical use regulations must be part of modern chemical education. Chemists need to know about the toxicological and hazardous properties of the chemicals with which they work. This knowledge is necessary in order to understand the relationship of these properties to chemical health and safety issues, governmental regulations arising from OSHA and other laws, and industrial hygiene in the chemical field. The basis of this knowledge can be developed in the standard chemistry lecture courses while maintaining the integrity of sound chemical education. When combined with other laboratory and occupational safety training the result will be a valued "OSHA Savvy Chemist" who can work responsibly with modern chemical science and technology. This Chapter addresses the role of the academic community at the college and university level in this shared educational effort.

There is a growing need in chemical education to address the basic concepts necessary for developing an understanding and acceptance of modern chemical health and safety (CH&S) issues by chemists and other scientists and engineers who handle chemicals. Those involved with chemistry and chemicals should have as some of their goals a societal acceptance of the manufacture and use of chemicals, and the impact of these chemicals on the environment. The attainment of these goals is based primarily upon the responsible attitudes and actions of those directly involved with chemistry, chemicals, and chemical education. Practitioners in related disciplines share these responsibilities. The regulatory laws of the Occupational Safety and Health Administration (OSHA) can serve as a focus for many current chemical health and safety issues. Those who have the education, training, and experience leading to responsible behavior in areas of chemical health and safety can be termed "OSHA Savvy." This Chapter will focus on the role of the academic community in educating the OSHA Savvy chemist by discussing the following areas:
    1. Characterizing the OSHA Savvy Chemist

Throughout This Chapter pertinent excerpts from *Prudent Practices in the Laboratory: Handling and Disposal of Chemicals* (1) are used to support and amplify certain points. This is referred to hereinafter as *Prudent Practices* 1995. The earlier editions (1981) of this volume (1) and a companion volume (1983) (1) served as the basis for the non-mandatory recommendations of the National Research Council for chemical hygiene in laboratories that appears as Appendix A in the OSHA Laboratory Standard 29 CFR 1910.1450. This volume is widely cited and accepted by the chemical laboratory community, and represents a basic reference for response to OSHA by academia. An excerpt from the discussion in Section 1.D, **Special Safety Considerations in Academic Laboratories,** expresses the basic the philosophy of this Symposium Volume.

"*Forming the foundation for a life-long attitude of safety consciousness, risk assessment, and prudent laboratory practice should be an integral part of every stage of scientific education - in the classroom, in textbooks, and in the laboratory from the earliest exposures in primary or secondary school through graduate and postdoctoral training. Teaching and academic institutions have this essential and unique responsibility.* (p. 16).

Appropriate references are also made to the topical chapters in *This Volume.*

Characterizing the OSHA Savvy Chemist

In the most general sense we would define the "OSHA Savvy" Chemist as an individual who has "bought in" to the "Safety Culture" (2), and who has an awareness of the regulations promulgated by the Occupational Safety and Health Administration, in particular the Laboratory Standard and the Hazard Communication Standard. This individual is able to assess the potential impact of these and related regulations on a particular chemical use. This includes all chemists, as well as those in other professional fields involving chemicals. These and some additional characteristics of the OSHA Savvy Chemist are listed in Table I. Additional characteristics would be needed to define more specifically the OSHA Savvy college chemistry professor, the OSHA Savvy academic research chemist, the OSHA Savvy undergraduate/graduate student, and other academicians (deans, administrators). These points will be developed throughout This Chapter.

The Need for OSHA Savvy Education. Is there really a need for the OSHA Savvy chemist? Many in the academic field may hold the view that this individual would be

Table I. Some Important Characteristics of an OSHA Savvy Chemist.

- Accepts safe behavior as a normal expectation
- Expects support of chemical health and safety policies from supervisors
- Is aware of the OHSA chemical hazard regulations and their potential impact
- Knows how to apply regulations and/or solicit help when needed
- Knows the basic principles of chemical safety, measurement, and assessment
- Can assist in creation and maintenance of safe practice in employment settings
- Behaves in a safe and responsible manner as a chemical professional
- Has an awareness of related requirements of other regulatory agencies
- Can apply chemical principles to risk assessment
- Uses prudent practices in planning new experiments and procedures
- Has an awareness of the chemical safety literature
- Renews safety training and learns new procedures
- Communicates the safety ethic to colleagues and the public

a specialist who has chosen to become a health and safety professional. While there has been the development of an infrastructure of such professionals, the thesis of This Volume is that the awareness and practice of chemical safety knowledge should be characteristic of all chemists. Although the focus is on OSHA rules and regulations to serve as points of reference, the scope of the concept is broad based. There have been such rapid changes in the area of chemical health and safety that many in the teaching and research field may be out of date. The need for chemical health and safety education runs the gamut from modifications of undergraduate chemistry courses to specialized professional training. Some reasons are listed in Table II. These will be discussed further in This Chapter.

Table II. Reasons for OSHA Savvy Education

1. Promulgation of the OSHA Laboratory Standard
2. Increased safety responsibilities and accountability assigned to academia
3. Need to write and follow chemical hygiene plans
4. Development of safer teaching methods such as microscale laboratories (4)
5. Development of more sensitive methods of detection of chemical hazards
6. Necessity of communication with health and safety professionals
7. Inconsistent safety consciousness in students entering college and graduate school
8. Increasingly litigious society
9. New hazardous chemical disposal requirements
10. Expectation of safety awareness in new employees
11. Development of new safer chemical technology such as Green Chemistry (5)
12. Need to combat chemophobia

And the list goes on. Several of these reasons are developed further in Chapter 11.

Apparent Regulatory Burden and Compliance.

Part of assuring the public of acceptable chemical health and safety is the development of a system of regulations affecting the workplace and the environment. The Hazard Communication Standard 29 CFR 1910.1200 promulgated under OSHA in 1970 was designed to minimize chemical exposure to individuals in an industrial production setting, and was not taken seriously by most academic chemists. The 1990 OSHA Laboratory Standard, 29 CFR 1910.1450, (6) specifically addressed the use of chemicals in academic and industrial laboratories, and brought the issue of compliance with chemical regulatory laws directly into the academic setting. Most academicians are now aware that certain requirements of OSHA apply to them, their graduate students, and any other persons under their supervision as employees. They also should recognize that they have certain responsibilities to assure that these persons receive applicable safety information and training. Individual chemistry teachers, researchers, or laboratory workers cannot be expected to know all of the OSHA and other federal, state, and local regulatory laws, but they need to be familiar with the general provisions and their own specific responsibilities. Many institutions have established Environmental and/or Occupational Health and Safety Offices, or have a staff of EH&S professionals to oversee compliance, and serve as a resource. Some larger institutions have such staff as part of their chemistry department (see Chapter 7), while in small institutions an individual faculty member may be assigned the position of institutional chemical hygiene officer. In any case there is an implied order of responsibility and associated administrative structure shown in Table III., using as an example the situation at the University of Wyoming.

Table III. Levels of Responsibility for Chemical Health and Safety in Academia

Health and Safety Hierarchy at the University of Wyoming

1. University President - Board of Trustees
2. Vice President for Finance
3. Head, Risk Management Office
4. Head, Environmental Health and Safety Office
5. Institutional Chemical Hygiene Officer
6. A&S College Safety Coordinator - Dean of Arts and Sciences
7. Chemistry Department Head
8. Chemistry Department Chemical Hygiene Coordinator
9. Faculty Teaching/Research Laboratory Supervisor
10. Staff Laboratory Coordinator
11. Post Doctoral Researcher
12. Graduate Teaching/Research Assistant
13. Undergraduate Laboratory Employee
14. Work Study Student
15. Student in Laboratory Courses[a]

[a]Not strictly covered by OSHA regulations, but covered by UW Trustee regulations.

There has been considerable discussion concerning the safety training of students with respect to OSHA. Is there any apparent or implied responsibility to treat them accordingly? *Prudent Practices* 1995 states:

" *OSHA standards apply only to 'employees' of laboratory facilities. In many cases, students are not employees within the scope of the Occupational Safety and health Act, but both moral and legal considerations suggest that colleges and universities provide the same protections to students as are provided to all employees regularly working in the laboratories.* " p. 204.

Similar statements are found in *Safety in Academic Chemistry Laboratories* (7). *'Academic institutions have a moral and professional responsibility to train students in safe laboratory practices.* " p. 41.

"Although nonemployee students are not covered by these (Chemical Hygiene Plan) regulations, each student should attest by signature that he or she has read and understands the safety rules. " p.38.

"-the reason for making safety an integral part of all laboratory operations is not the regulations and their associated legal implications, but rather that doing things safety is <u>the right way to work.</u> p. 2.

Some institutions for taken the step of stating that their Chemical Hygiene Plan applies to students. Although OSHA would not be concerned with the implementation of such an extension, any lawyer representing a student injured in an academic laboratory experiment would hold the faculty supervisor (and all related supervisors and administrators) of that student responsible for following the institutional rules. Therefore it is vital that if the CHP of an institution has been extended to students, all called-for training and other aspects be carried out as stated for employees. However, this requirement could present problems due to the time and staff required, and chances are high that the students may not receive the same information and training. This would result in a potential legal liability. A more desirable alternative may be to implement a specific safety program for students, such as described in Chapter 5. Other universities have developed similar student safety policies (8)

Chemical Health and Safety Education: A Shared Responsibility

The question arises as to how a chemist is to acquire the education needed to become OSHA Savvy. Education is a life-long activity. Chemists who lack OSHA Savvy can become OSHA Savvy anytime in their career, but modern needs dictate that this time should be as soon as possible . The two major sets of OSHA regulations that chemists need to work with are the original Hazard Communication Standard and the more recent Laboratory Standard. Achieving the goals listed in Table I. requires a coordinated experience of chemical education and specific safety training. Academia is not regulated to change basic chemical education, and OSHA regulations apply strictly to employer-employee situations. However, academicians who hold the view that most of the goals amount to on-the-job training, and are the responsibility of the employer, may not appreciate the fact that those hiring chemists or individuals expected to have some knowledge of chemistry are assuming some fundamental chemical safety knowledge as part of a basic understanding of chemicals. Also, they

8

are depriving students of the opportunity to "buy-in" to a culture of safety. Even though the regulations themselves are addressed to employment situations (unless modified as discussed above), individuals educated in an environment where there is not a culture of safety will not develop the behavior patterns necessary to apply on-the-job training effectively.

Since an administrative structure like that shown in Table II is required for employee safety at academic institutions, it is reasonable to take advantage of it in the educational process. Chapter 11 presents an example of integrating a chemical hygiene plan into the undergraduate curriculum. Although it may appear that many OSHA rules and regulations are inflexible, the concept of chemical use standards under the Laboratory Standard are "performance based." This means that within certain limits some allowance is provided for novel approaches to safe practice as long as behavior indicates proper attention to safety issues. For example, the flexibility of performance based standards allow chemical hygiene plans to be adapted to the procedures of individual laboratories. An example of this approach is discussed in Chapter 9. There are certain issues that cannot be neglected, including training, keeping MSDSs, labeling requirements, and providing necessary protective equipment, but these issues ultimately reflect the access to knowledge and understanding of chemical safety.

" Chemicals are handled not only by chemists, but also by biologists, physicists, geologists, materials engineers, and others." p.1.

Responsibility in the Chemistry Department to Safety Education. There are many ways to achieve a basic departmental ethos that emphasizes safety. Consistent use of policies and programs listed in Table IV. Additional ideas, such as safety messages on bulletin boards, posting MSDSs, hosting seminars, and using appropriate signage are discussed in Chapter 11. These policies present a clear message to students, faculty, and staff that chemical safety is a normal part of working and learning in a chemistry department.

Table IV. Some Basic Policies for Departmental Safety Programs

1. Establish Departmental Safety Committees that Include Students
2. Develop a Departmental Chemical Hygiene Plan
3. Designate a Departmental Chemical Hygiene Officer
4. Establish In-House Safety Inspections
5. Require Safety Training for All Laboratory Faculty and Teaching Assistants
6. Require Eye-Protection for All Persons in Laboratories
7. Cooperate Fully with College/University EH&S Office
8. Ensure Accessibility of MSDSs received in the Department
9. Make Available Standard Safety Reference Books

Several examples of academic chemical hygiene plans are presented in Part II, Chapters 6 - 10 of *This Volume.*
"The committee (the NRC-National Research Council-Committee that prepared "Prudent Practices") *recommends that chemical laboratories establish their own safety groups or committees at the department level, composed of a cross section of*

laboratory workers, including students and support staff as well as faculty in academic laboratories." p.11

"A departmental safety group must have the full support of the institution's administration." p.11.

Safety in Academic Chemistry Laboratories. Most safety instruction received by students in chemistry courses comes from the laboratory portions of a college or university general chemistry courses. For many this is their first experience in safety education. The academic community has a good record in this area. *Safety in Academic Chemistry Laboratories,(7)* now in its 6th Edition, was initiated in 1971, well before OSHA. This booklet has been a standard and well accepted source for many years. Other guidelines have been published in the Journal of Chemical Education (3). However, instruction in laboratory courses is usually given by teaching assistants, and the quality of instruction may vary. Many incoming graduate students are not OSHA Savvy, and may have had their first real safety education a week before their first laboratory session. There are particularly severe deficiencies in new T.A.s from foreign countries where safety policies may be non-existent. Students often view this type of instruction as related only to the laboratory operations, and do not consider it as applicable to other situations, or as part of their basic chemical education. Reference to OSHA is usually lacking.

Since the general chemistry laboratory safety training and experience is so often the starting point for all subsequent safety education, it is vital that it be presented as completely and professionally as possible. Bad habits learned by or poor attitudes demonstrated to students at this time lay an unstable foundation for the eventual development of a sound safety behavior. Some important factors to consider in the general chemistry and subsequent laboratories are shown in Table V. Consideration of items 2. and 3. will impress students with the seriousness of safety issues. EH&S staff may be able to assist in training students, but it should be realized that their first responsibility is for mandated safety training called for by OSHA. The faculty lecturer should emphasize that disregarding safety rules and regulations can lead to failure in the laboratory and in most cases corresponding failure in the lecture course. Since the faculty member in charge of any laboratory course would probably be considered the individual most responsible in case of legal action resulting from an accident involving a student, it would seem ill-advised to place all the responsibility for the safety training of those students in the hands of a neophyte teaching assistant. It is certainly the responsibility of the faculty to monitor the safety training, and better still be involved with at least part of it.

Items 13.-19. are particularly important factors in introducing safety issues into the curriculum. These areas can be expanded into the lecture portion of the course, as discussed below. Other chapters in *This Volume* treat specific examples of integrating safety issues. Chapter 12 discusses the educational value of a labeling system. Chapter 13 gives examples of correlating MSDS information from a laboratory exercise into the lecture material. The importance of testing on safety information is the topic of Chapter 15.

Table V. Safety Considerations for Undergraduate Laboratories

1. Proper Safety Training for All Laboratory Instructors and Teaching Assistants
2. Faculty Course Lecturer Supervision of Student Lab Safety Training
3. EH&S Personnel Participation in Initial Student Lab Safety Training
4. Chemical Splash Goggles for All General and Organic Laboratories
5. Appropriate Eye Protection for All other Laboratories
6. Adequate Hood, Eyewash, and Shower Facilities
7. General Laboratory SOPs for Experimental Control
8. Specific Safety Concerns Presented in Writing for all Experiments
9. Introduction of Safety Terminology and Acronyms
10. Waste Disposal Protocols
11. Stockroom Security Proper
12. MSDS Availability for All Chemicals in Use
13. Labels for Ready-Use Containers
14. Pre-Lab Exercises Requiring Use of MSDSs and other Safety References
15. Safety Topics Discussed in Lab Reports
16. Safety Points for all Lab Experiment Grades
17. Safety Questions on Lab Quizzes and Exams
18. Safety Policies Continued in all other Laboratory Courses
19. Additional Safety Issues Introduced in Advanced Laboratory Courses

The following two statements from *Prudent Practices* 1995 support some of the considerations of Table V.

"Future (laboratory) manuals should include questions and assignments that involve the student actively in considering the risks, regulations, and waste disposal costs for alternative approaches to the problem under discussion." (p.3.)

"The committee recommends that any laboratory using hazardous chemicals should provide appropriate training in safety and waste management for all laboratory workers, including students in laboratory classes." (p.11.)

Safety in the Lecture. The formal OSHA requirements for safety training of employees, including faculty, and the realization that the implementation of basic laboratory safety policies for students represent moral, professional, and legal responsibilities of faculty have resulted in at least minimal chemistry laboratory safety programs at most academic institutions. This is part of the *training* (9) of the OSHA Savvy Chemist However, the goal of *educating* the OSHA Savvy Chemist requires attention in the basic lecture curriculum. This is the major thrust of This Volume. In *Undergraduate Professional Education in Chemistry: Guidelines and Evaluation Procedures* (10), the ACS Committee on Professional Training (CPT) states that "Discussions of current health and safety issues must be an integral part of the chemistry curriculum, beginning early in the core courses with discussions of toxic effects of chemicals, their flammability, their explosive character, and in some instances their radiation hazards. Recognized safety practices should be stressed both in lecture and laboratory discussions, including but not limited to compliance with the regulations of OSHA, the recommendations in the ACS manual *Safety in Academic Chemistry Laboratories* and in NAS-NRC *Prudent Practices in the Laboratory*, and applicable state and local regulations. Students should be knowledgeable regarding the physical, chemical, and biological properties of the substances they handle. They should recognize hazards and be prepared for worst possible situations." (11)

In spite of the recommendations of the ACS CPT and those in *Prudent Practices*, almost no beginning chemistry textbooks treat the subject at all, and laboratory manuals usually cover only basic laboratory safety. Consequently, few of those teaching chemistry courses have chosen to integrate chemical health and safety issues in the traditional college and university chemistry lecture courses. Some textbooks have added a chapter on toxic chemicals (12) Since a college or university course in General Chemistry is usually a requirement in any discipline where chemistry might become part of the professional activities, this should be the starting point for the introduction of OSHA related material.

When chemistry instructors who are not yet OSHA Savvy read the recommendations of the CPT and *Prudent Practices* 1995, or are challenged in some other way to treat safety issues, such as reading *This Volume*, some typical responses and concerns are shown in Table VI.

Table VI. Questions and Concerns about Safety in the Curriculum

1. The topic is not treated in my textbook.
2. There is no time for this topic.
3. The topic is inappropriate for my course in (subject) chemistry.
4. There is no source of reference material

Responses to Questions and Concerns. Incorporation chemical health and safety issues into the undergraduate curriculum will take a concerted effort of all parties involved. The effort should be made at several points, but the best starting point is in general chemistry. Additional suggestions are given below.

General Chemistry. Chapter 14 of This Volume offers an outline for an introductory chapter for general chemistry texts. It addresses chemical issues related to industrial hygiene. The concept involves presenting general definitions of terminology that are important in understanding the hazardous properties of chemicals, followed by problems using these terms The argument is made that students should begin to learn about hazardous properties of chemicals along with their other physical and chemical properties. Even if some of the terminology may be covered in the corresponding laboratory course, the additional in-depth treatment that can be presented in the lecture, accompanied by appropriate related problems and testing, helps students to see the basic importance of the subject. Other terms and information may be covered only in the lecture. The recommendation made here is that instructors develop such an introductory CH&S chapter appropriate to their own course. Some topics to consider are shown in Table VII. This list is expanded in Chapter 14. Continued use of appropriate chemical health and safety concepts throughout the general chemistry course and in advanced courses can be based on this introduction.

It is usually possible to substitute CH&S related problems for others that might represent rote exercises. With proper selection, the incorporation of safety education can be done without the requirement of additional time, or the displacement of necessary topics. Most instructors spend some time in their courses presenting applications of the basic principles, or make some choices among optional material in the text. For example, references to environmental issues are often made. Some of these examples are in fact health and safety issues as well, and one can kill two birds with one stone with little extra effort.

Table VII. Chemical Health and Safety Topics for General Chemistry Courses

1. Elementary Toxicology: Routes of entry, target organs, health effects
2. Hazard Terminology: Health hazards, physical hazards, hazard warnings (13).
3. Important Acronyms: TLV, LD_{50}, (13)
4. Stability, Reactivity, Incompatibility
5. Awareness of Governmental Agencies: OSHA, EPA,
6. Awareness of Non-Governmental Agencies: ACGIH, NIOSH, NTP
7. Awareness of Important Laws and Regulations: OSHA, EPA, RCRA, CERCLA,
8. Appropriate Problems: Unit conversions, TLV, LD50 calculations, reactivity and incompatibility reactions, experimental planning.
9. Appropriate Testing: Lab quizzes, lecture quizzes, lecture exams.
10. Sources of Information: Labels, MSDSs, safety data sheets, *Prudent Practices*

Organic Chemistry. Many students who do not continue as chemistry majors, but for whom chemistry will be an important professional area, take organic chemistry. Therefore this course is usually the last organized experience in chemistry, and it is important that the chemical health and safety education learned in general chemistry be supported and expanded. Laboratory safety techniques are usually still well supported, and the T.A.s are often more experienced. However, incorporation of CH&S topics into the lecture material is even less common than in general chemistry. On the other hand, some laboratory manuals address certain new health and safety topics. *Microscale Organic Laboratory* by Mayo, Pike, and Butcher (14) devotes a chapter to "Safety and the Environment of the Laboratory". There is a discussion of TLV and TWA, and methods of estimating vapor concentrations in the laboratory are presented. *Macroscale and Microscale Organic Experiments* by Williamson (15) discusses the concept of reducing hazardous wastes from laboratory experiments by making the treatment steps part of the written experimental procedure, and includes such steps in several syntheses. This approach generally is exempt from a RCRA permit (16). It has become recommended for procedures submitted to *Organic Syntheses* and *Inorganic Syntheses*, and the experimental parts of articles submitted to ACS chemical journals. All chemistry students should be made aware of this approach for their future course work and professional activities. The chemistry of these steps could be presented in the lecture course.

Resource Material for Other Courses. There are numerous opportunities for the introduction of alternative problems and laboratory exercises in other standard chemistry courses, or for supplementary lecture material. Instructors are challenged to develop their own contributions to the education of the OSHA Savvy Chemist. While a single definitive source remains to be written, there are many sources of information that could be useful. A few possibilities are given below:

Substitutes for Hazardous Chemicals in the Workplace (13). Although written from an industrial point of view, this volume contains information useful for modifying laboratory procedures in both undergraduate courses and research.

Chemical Health & Safety (17). This fairly new bimonthly journal published by the ACS Division of Chemical Health and Safety (CHAS). There are usually articles which can be used to supplement all chemistry courses. Those teaching chemistry courses should be aware of this resource.

Bretherick's Handbook of Reactive Chemical Hazards (18) is a basic reference that can be used to find supplementary chemical reactions for a variety of courses.

Environmental Chemicals Data and Information Network (ECDIN) (19). This is an Internet accessible factual data bank based in Europe containing a wide variety of environmental and CH&S information. Much of the information is similar to that found on MSDSs, but unlike those extensive references are given. Additional information is available that is not found on MSDSs. Two categories that are useful in developing supplementary lecture material include Analytical Methods for Detection and Chemical Processes used for production.

Definitions, Conversions, and Calculations for Occupational Safety and Health Professionals (20). This volume should be very useful for designing alternate problems for general and analytical chemistry courses.

Stability and Reactivity data from MSDSs. These data offer a wealth of information for predicting or interpreting chemical reactions that would be applicable to various courses, including general chemistry, organic chemistry, and inorganic chemistry. In particular, MSDSs from MDL Information Systems, Inc. list extensive reaction information under "Incompatibilities." For example, magnesium is incompatible with barium carbonate because of the formation of an explosive acetylide (21). Have your students write the reaction.

Safety Education in Academic Research Laboratories

Undergraduate Research. Most chemistry majors engage in some sort of undergraduate research activity. This may be a student's first experience involving hazardous materials for which established safety procedures have not been established. New compounds may be prepared that have unknown hazardous properties. At the University of Wyoming, all students enrolling for undergraduate research in Chemistry must attend the same safety training sessions required for new employees (graduate students and post-doctorals) as a required part of the course. In any case new undergraduate research students must be given some sort of initial safety orientation. In addition, it is particularly important that these students receive careful oversight and leadership from the faculty supervisor. If previous safety education and training has been lacking, these students may be particularly vulnerable to laboratory accidents.

Graduate Research. Most new graduate students in chemistry departments enter with graduate assistantships and are classed as employees. In this category they should receive whatever OSHA required information and training the institution has developed under its chemical hygiene plan. However, with both graduate students and undergraduate research students, it is important to consider the following, from *Prudent Practices* 1995.

"Formal safety education for advanced students and laboratory workers should be made as relevant to their work activities as possible. Training that is conducted simply to satisfy regulatory requirements tends to subordinate the relevant safety issues to details associated with compliance." (p. 17)

What is important is the professionalism demonstrated by the teacher.

"When the principal investigator offers leadership that demonstrates a deep concern for safety, the university safety program thrives. However, if the principal investigator's attitude is laissez-faire or hostile to the university safety program, careless attitudes can take hold of the whole group, and set the stage for accidents, costly litigation, and expensive reeducation for those who move on to a more responsible institution." (p.18)

The Benefits of OSHA Savvy

In addition to the basic benefits discussed related to doing the right thing and not being victimized by ignorance of the laws, the OSHA savvy academic chemist can derive additional benefits. Whether through the presentation of formal courses or by mentorship as a research advisor, the academician is basically an educator. For those involved in course development and improvement, areas of traditional education can be augmented. Explaining these aspects to the students involved adds to their basic education as well. For example, pollution prevention has become a major thrust of the chemical industry, and those learning chemistry now should be aware of this trend. Three major areas involving source reduction, treatment, and reuse and recycling could be incorporated into laboratory courses in the following ways:

- Develop modified experiments to use less hazardous starting chemicals, or produce less hazardous waste.
- Incorporate treatment steps into existing experiments to reduce the hazard of waste products.
- Use products of one procedure as the staring materials for the next.

Students could be directly involved by incorporating these exercises into special projects or undergraduate research.

Some additional benefits to consider are these:

- Minimize the risk of adverse publicity from chemical accidents/incidents for the academic institution and the individual.
- Build public confidence in academic chemistry and chemical technology.
- Participate in the process, educate the regulators through professional society activities, explain the effect of a regulation on academic chemistry.
- Get involved in public comment, public hearings on proposed legislation.
- Be able to respond to safety requirements of grant proposals.
- For those entering industry, minimize the need for reeducation on safety matters.
- Provide the basis for judgment and modification with new CH&S situations, or decision making in a new employment role.
- Develop basic research projects related to safer chemical processes.

Recommendations for Future Developments

The dilemma that exists within the academic community in regard to making chemical health and safety more of an integral part of the curriculum reflects a lack of leadership in both the academic and industrial areas. Some recommendations for increasing the effectiveness of CH&S educational efforts are given below.

1. *Safety in Academic Chemistry Laboratories* (7) should be required for all students taking college and university general chemistry courses. Reference should be made to this booklet as part of the laboratory safety training. Its use should be reinforced in other chemistry laboratory courses. Students who take only general chemistry should be advised to keep the book as a reference in other laboratories they may take where chemicals are used . They should also be advised that the book represents basic laboratory safety practices that are also applicable to employment in chemically related laboratory jobs.

2. All those involved with laboratory chemical education and research should own or have ready access to P*rudent Practices in the Laboratory: Handling and Disposal of Chemicals* . Section **1.D** should be required reading for all college and university chemistry faculty. **This section presents a convincing distinction between safety education and safety training.**

3. CH&S topics need to be integrated into the lecture portion of general chemistry courses, and other higher level chemistry courses including organic, quantitative

analysis, physical chemistry, and others. These topics need to be related to the basic chemical principles usually covered in the particular course. PART III of *This Volume* addresses several methods for doing this.

4. As an incentive to address CH&S issues, the ACS Committee on Professional Training should evaluate coverage of safety topics called for in the CPT Guidelines as part of the five-year evaluations of ACS accredited chemistry programs, and make note of deficiencies as they do in other basic areas. Without this the CPT recommendations will continue to be ignored.

5. Regular hour exams, quizzes, and ACS Standardized Tests in general chemistry and other areas should include more CH&S questions as reflected by the CPT recommendations. Chapter 15 of *This Volume* addresses the general concept of testing.

6. Chemistry textbook authors and chemistry instructors need some leadership in regard to CH&S topics to incorporate into books. In this regard there needs to be a body of recommendations about these topics. These recommendations could come from a committee or task force sponsored by the ACS that includes representatives from:
   a. The Committee on Professional Training
   b. The Committee on Chemical Safety
   c. The Division of Chemical Health and Safety (CHAS)
   d. The Division of Chemical Education (CHED)
   e. Major textbook authors
   f. Major textbook publishers
   g. Chemical employers
   h. EH&S Professionals
A set of these recommendations about CH&S topics, terms, and concepts that should be incorporated into undergraduate courses and textbooks should be published in *J. Chem. Ed.* and *Chem. Health and Safety*. Alternatively, a booklet similar to "Safety in Academic Chemistry Laboratories" could be published.

7. Employers should ask about CH&S awareness as part of job interviews. This practice will be useful both to chemical educators and to employers. The employers can evaluate the general level of CH&S knowledge in new graduates in chemistry and chemistry supported fields, and note strengths and weaknesses. This information would need to be transmitted back to the academic community.

8. New employees in non-academic laboratories and the chemical industry should be made aware of their rights, responsibilities, and expected extent of knowledge in regard to CH&S related to the Laboratory Standard and/or Hazard Communication Standard immediately on starting the job. They should be told what education and training they can expect from their employer. With this evaluation new employees will have some idea of what they need to study on their own.

9. CH&S Representatives from the Chemical Industry should offer to speak before Chemistry Department seminars, student affiliate groups, and at national meetings.

Ultimately, we hope for inclusion of these topics into main-stream chemistry courses, textbooks, and lab manuals. Authors, editors, and publishers of these volumes need to be convinced that the time has come for the "culture of safety", already well accepted by most other areas of the chemical profession, governmental agencies, and the public, to be extended to the level of basic chemical education.

Conclusion

It is fitting to conclude this discussion with the following excerpt from *Prudent Practices* 1995:
"Public support for the laboratory use of chemicals depends on compliance with regulatory laws as a joint responsibility of everyone who handles or makes decisions about chemicals, from shipping and receiving clerks to laboratory workers and managers, environmental health and safety staff, and institutional administrators. This shared responsibility is now a fact of laboratory work as inexorable as the properties of the chemicals that are being handled. The use of chemicals, like the use of automobiles or electricity, involves some irreducible risks. However, all three of these servants to humankind have demonstrated benefits that enormously outweigh their costs if they are handled sensibly." (p. ix)

Literature Cited.

(1)   (a) *Prudent Practices in the Laboratory: Handling and Disposal of Chemicals;* National Academy Press: Washington, D.C., 1995. (b) *Prudent Practices for Handling Hazardous Chemicals in Laboratories;* National Academy Press: Washington, D.C., 1981. (c) *Prudent Practices for Disposal of Chemicals from Laboratories;* National Academy Press: Washington, D.C., 1983.

(2)   ref. 1a., Ch. 1.

(3)   *Safety in the Chemical Laboratory;* 4 vols.: Reprints from *J. Chem. Educ.,* Journal of Chemical Education; Easton, PA, 1964-1980.

(4)   Williamson, K. L. *Macroscale and Microscale Organic Experiments, 2nd Ed.;* D. C. Heath and Co.: Lexington, MA, 1994.

(5)   *Designing Safer Chemicals: Green Chemistry for Pollution Prevention; DeVito, S. C.; Garrett, R. L,* Eds.; ACS Symposium Series No. 640, American Chemical Society, Washington, D.C., 1996.

(6)   ref. 1a., Appendix A: pp. 219-225.

(7)   *Safety in Academic Chemistry Laboratories, 6th Ed.,* ACS Committee on Chemical Safety; American Chemical Society: Washington, D.C., 1995.

18

(8)   (a) **http: //safety.uwyo.edu/risk/manual/studentsa.pdf** (b)
        http://www.uwyo.edu/A&S/chem/safety2.htm
(9)   ref. 1a., p. 14, footnote 2.
(10)   *Undergraduate Professional Education in Chemistry: Guidelines and
        Evaluation Procedures,* Committee on Professional Training, American
        Chemical Society: Washington, D.C., 1992.
(11)   ref. 11., p.
(12)   (a). Joesten, M. D.; Wood, J. L. *World of Chemistry;* Saunders College
        Publishing: Orlando, FL, 1996; Ch. 18. (b) Snyder, C. H. *The Extraordinary
        Chemistry of Ordinary Things, 2nd Ed.;* Wiley: New York, NY, 1995; Ch. 18.
(13)   Goldschmidt, G. *Substitutes for Hazardous Chemicals in the Workplace;* CRC
        Press, Boca Raton, FL, 1996.
(14)   Mayo, D. W.; Pike, R. M.; Butcher, S. S. *Microscale Organic Laboratory,
        2nd. Ed.;* Wiley: New York, NY, 1995.
(15)   ref. 4., p. vi.
(16)   ref. 1a.,
(17)   *Chem. Health Safe.* **1994-1997,** *1-4.*
(18)   Bretherick, L. *Bretherick's Handbook of Reactive Chemical Hazards;* 5th ed.;
        Urben, P. G., Ed. Butterworths: London, 1996.
(19)   **http://ulisse.ei.jrc.it/Ecdin/Ecdin.html**
(20)   Finucane, E. W.; *Definitions, Conversions, and Calculations for
        Occupational Safety and Health Professionals;* Lewis Publishers, CRC Press,
        Boca Raton, FL,1997.
(21)   MSDS# OHS13290, MDL Information Systems, Inc. 14600 Catalina St., San
        Leandro, CA, 94577

Chapter 2

Safety Is Part of the Job: Why Employers Want Workplace-Savvy Chemists

Andrea Sutherland

Andrea Sutherland Consultants, 1143 Webster Street, Palo Alto, CA 94301–3246

Universities and colleges train scholars. Chemical education doesn't always match the pragmatic application of scholarly pursuit in the workplace.Today's graduate now works in a highly regulated workplace and few have been prepared to step from an academic lab to an industrial lab.

Consider the university setting - students learn chemistry on a microscale, lab experiments use less hazardous materials than a commercial lab. The diversity of chemistry in the workplace setting can't be matched in classrooms.

Today's business environment requires skills, problem solving, technique and ingenuity. Universities cultivate these skills, but most students don't learn about chemical disposal, emergency response or liability.

New graduates entering a chemically related workplace are often ill-prepared to face industrial chemical health and safety requirements. This is particularly true in the new technologies and discovery chemistry workplaces. New graduates are attracted to biotechnology and chemistry related high technology. This workplace is very dynamic. The goal is to discover new pharmaceutical products or devices to treat and diagnosis disease. Discover chemistry is evolving and the creation of thousands of new molecules is the norm. (These are the fields in which the author has had the most experience with safety issues). The reasons for this lack of safety preparation reflect certain aspects of the academic world, which result from safety and economic considerations. Academic education emphasizes general principles, along with training in certain traditional techniques. Laboratory procedures are often designed to use small amounts of low-hazard chemicals. Reagents are usually prepared by teaching assistants, and students are not usually involved in waste disposal. In many cases, instrumentation for teaching labs is traditional and antiquated. Students in basic courses can be unaware of the sophisticated instruments they may encounter on the job.
  Many chemists involved in college and university teaching and research have spent their whole career in academia and have had no experience with or understanding of the modern safety requirements of the industrial world. For example, most are unfamiliar with the use of the MSDS (Material Safety Data Sheets) and do not read them. The requirements of OSHA (Occupational Safety Health Act) and other regulatory agencies impacted the non-academic workplace

sooner, and have been taken more seriously. (1) Most institutions now recognize that these workplace rules and regulations apply to professors, teaching assistants, graduate students, and other academic employees (5), but they are often not enforced as strictly as in the non-academic workplace. Unfortunately, if not practiced by their mentors, the students can be deprived of being introduced to and seeing the application of safety principles. Nevertheless, the safety seed needs to be planted early and cultivated regularly. The academic laboratories are the place to start. Most first year general chemistry courses provide some of the elementary safety policies(6) related to the laboratory, and eye protection is almost universally required. Discussion of additional safety issues usually are not treated in subsequent courses. It is important at least to continue the basic safety policies of beginning courses.

The major reasons that employers want safety-savvy chemists, (or other employees who wear a chemist's hat in their workplaces), reflect basic economic concerns. The cost of hiring a chemist or other technical graduate can exceed $100,000. (2) Any time lost to accidents represent a loss to the company. An individual with poor safety habits is a potential liability to him/herself, to co-workers, and to the company and its facilities. Such individuals who move on to supervisory positions can pass on bad habits to others. In addition to accidents and associated litigation, the possibility of regulatory fines pose another concern.

The areas where safety deficiencies, in new biotechnology employees, are most apparent include the following:

* Chemical Toxicity

* Chemical Handling

* Emergency Response

* Safety Literature and Communication Skills

These deficiencies are probably similar in other fields of the chemical industry. This chapter looks at some differences in the academic teaching environment and the industrial workplace that lead to these problems, explains why they are important, and offers some possible solutions.

Chemical Toxicity. In chemistry courses students learn a lot about what a chemical **can do for** them, but they know woefully little about what a chemical **can do to** them[4]. This deficiency in their chemical education is a result of standard and currently recommended academic practices. (For example, almost all new laboratory manuals in general and organic chemistry have eliminated the use of benzene and dichromate because of their carcinogenic status.) For reasons of safety and economy teaching laboratories tend to make use of small amounts of reagents with minimum toxicity, and use low-risk procedures, such as microscale. Students have only a single or at most a few exposures to any one chemical, and learning about a chemical's toxicity is minimal. These procedures often continue into advanced courses, and even research projects. Spills and waste disposal are handled by the instructors.

In the workplace, this lack of understanding of chemical toxicity can result in serious problems. New graduates need to realize the use of toxic chemicals cannot be avoided. Large quantities of chemicals are often used on a day-to-day basis, and it will usually be the new chemist's responsibility to handle waste disposal. They must know how to respond to a spill, and the proper personal protective equipment to use. Lack of understanding of chemical toxicity is related to the failure to make the proper choice of personal protective equipment, and to use it consistently. A new chemist in the workplace may make choices based on the less toxic materials used in classroom setting.

Chemical Handling, Labeling, and Waste Disposal. The new chemist in the workplace will usually be involved in one or more activities related to ordering, storing, labeling, and waste disposal of chemicals. Undergraduates rarely get any experience in these activities, since they are handled by the course instructor and laboratory assistants. Current academic practice limits storage of chemicals in teaching laboratories to those used for the particular laboratory session. Acids and bases, and other incompatible chemicals may be put in the same hood for use in an experiment, and students can get incorrect ideas about storage. Often they do not see actual chemical reagent labels, since reagents for the experiment are already prepared and have only the minimum labeling required for transferred chemicals. Some bad habits can develop, such as leaving a waste container open. This is a common practice during an academic teaching laboratory session, since the container is closed by the T.A. at the end of the lab. It may come as a surprise for new employees to see how much waste can be generated in a day. The workplace chemist needs to learn about ordering in proper quantities, compatible storage, acceptable labeling of hazardous chemical intermediates, and compliant waste disposal procedures. (2, 7, 8)

Emergency Response. Students rarely get any type of training in this area, other than to evacuate the teaching laboratory at the request of the instructor or T.A. Accidents do happen, when they do, the workplace safety personnel will not be standing by and may take some time for them to respond. A chemist often has to make initial choices in case of an emergency after calling for assistance. This could include use of a fire extinguisher, containment of a spill, application of first aid, or evacuation procedures. This is an area for which specialized on-the-job training will be required if employees are expected to be able to use devices such as self continued breathing apparatus (SCBA)(6). However, a new chemist in the workplace should be expected to understand the possible effects of chemicals on the senses and the ability to respond rapidly to an emergency, and not to expose themselves by unwise actions. Advanced planning for emergency response is vital to minimize losses.(2)

Safety Literature and Communication Skills. Life-long learning is important to success in any career. New graduates entering the chemical field may have to continue their education to make up for deficiencies in their chemical health and safety knowledge. A good working safety vocabulary needs to be developed. There is a large variety of safety-related resource materials including scientific articles, specialized journals, conferences, and technical documents such as material safety data sheets (MSDSs). MSDSs are designed for the workplace and chemists need to learn how to read and use them. Increased awareness of the safety information contained on hazard labels, beyond the recognition of pictograms, is necessary in the workplace.

For other workplace safety issues basic communication skills will be important. Employees are expected to be able to present their work at laboratory meetings, conferences, and in publications, and these can have a safety component. New employees are often assigned to safety committees, and will have to deliver oral and written reports. They may eventually be responsible for training others under their supervision. In emergencies, good English skills are essential for the rapid dissemination of instructions.

Recommendations. There will always be certain differences between the academic learning environment and the industrial workplace, and education has to continue on-the-job. However, there are steps that need to be taken on both sides to reduce the academic/industrial safety culture shock.

Academic Side. <u>Develop the Academic Safety Culture.</u> Probably the most important aspect of educating students about safety issues is for them to see a good working program in operation. The safety program must be accepted by the top administration and complied with throughout the institution. Professors need to be accountable for poor safety practices. Even though workplace rules and practices will be different, if students have been educated within an academic safety culture good habits will be developed, and they will be receptive to the workplace safety culture.

Incorporate Safety Topics into the Curriculum. An effective academic safety culture will extend safety issues beyond their elementary application in the general chemistry laboratory. Elementary safety vocabulary, elementary toxicology should be incorporated into the general chemistry course. Applicable safety policies should be continued and expanded in advanced courses, including graduate courses.

Additional Safety Education Opportunities. In some cases it may be appropriate to offer a separate safety course. Safety topics can be assigned in seminars. ACS Student Affiliate groups can get involved in safety activities.

Industrial Side. Address Safety Skills in Job Interviews. If job applicants come to expect some evaluation of their safety skills, then they will be more receptive to these issues in their education. In viewing equally qualified canidates, previous experience with workplace safety may be the deciding edge.

Inform New Employees of Company Safety Policies and Expectations. Before new employees start their job activities, they should be informed as fully as possible about what safety information and attitudes the employer expects them to have, and the overall company safety policies. They should understand at the beginning the importance of these issues in their success and advancement. This will allow them to evaluate the deficiencies in their safety education, and start to remedy them.

Make Industrial Internships More Available. Prospective employees with experience through internships fare better in the workplace. If more of these were available and well publicized more students could experience the workplace. These students could then discuss safety issues with others.

Offer Speakers on Workplace Safety Issues. Departments are always looking for seminar speakers. Individuals who are effective in explaining the importance of being OSHA-savvy could be effective in delivering talks to undergraduate courses, seminars, or student affiliate groups.

Offer Advisors to Academic Programs. Institutions or departments developing safety programs could benefit from the advice of industrial safety consultants.

Conclusion. The workplace is where most chemists arrive after graduation. A "work place savvy chemist" can be accomplished by industry and professional groups clearly stating their needs and expectations. Students are the consumers of education and should ask for an education that prepares them as completely as possible for the workplace and success. Universities and college's contribute is to lay the best foundation for their scholar/employees. There is continuity of purpose amongst the academic and industrial groups; and from my vantage, collaboration, innovation and competition are fundamental strengths for success.

References

1.   29 C.F.R. § 1910.120 (1990) Laboratory Standard, hazardous waste operations, warning labels, emergency response, employee training requirements, Chemical Hygiene Plan requirements
2.   Personal communication with Dr. Eric Gordon, Chief Scientific Officer, Versicor Inc., Fremont, CA. The cost is for workspace, equipment, and overhead. It does not include salary, benefits or supplies.
3.   42 U.S.C.§ 11022 (Emergency Planning and Community Right To Know Act) 1986
4.   Worker Health and Safety (industrial hygiene) is regulated by the Federal Government under the Williams-Steiger Occupational Safety and Health Act of 1970 (29 U.S.C. §§ 651-678 (1988)

5. Standards are promulgated by the Secretary of Labor, and may be based on information from interested persons, organizations of employers or employees, nationally recognized standards producing organizations, the Secretary of Health and Human Services, the National Institute of Occupational Safety and Health (NIOSH), a State, or the Department of Labor. 29 U.S.C. §655 (1988) The American Conference of Governmental Industrial Hygienists (ACGIH) and the American National Standards Institute (ANSI) are involved in the standard setting process. 29 C.F.R. §§ 1910.31, 1910.1500 (1990)

6. 29 CFR § 1910.1450 (f) (1990) Training must include methods that can be used to detect the release of a chemical, the hazards associated with the chemical and the information available to employees in the Chemical Hygiene Plan.

7. 29 CFR §§ 1910.133 (1990) (eye and face protection), 1910.134 (1990) (respirators).

8. 33 U.S.C. §§ 1251, 1311, 1314(b), (c), 1317 (b) (c), 1326(b) (1988) Discharges of wastes to the sanitary sewer are regulated under the Federal Water Pollution Control Act, as amended by the Clean Water Act.

9. Resource Conservation and Recovery Act (RCRA), 42 U.S.C. § 6921 (d), 6822 (a)(2)(3)(4)(5)(6)1988 Regulations regarding record keeping, labeling, use of containers, preparation and tracking of manifests, reporting, and waste minimization.

10. 42 U.S.C. §§ 7401-7642 (1988) The federal Clean Air Act as well as state and local requirements regulate emissions into the air.

Chapter 3

Laboratory Safety and New Jersey Institute of Technology's Chemical Hygiene Plan

Norman J. Van Houten

Department of Health and Environmental Safety, New Jersey Institute
of Technology, Newark, NJ 07102

The Chemical Hygiene Plan at NJIT relies on a series of compliance-oriented documents that ensure training, SOP development and assessment of safety in particular laboratory operations. The overall safety process relies on documented training and committee review of safe behavior. Both teaching and research laboratories are covered by the NJIT CHP. This paper describes general SOPs, training requirements and the process for developing and implementing specific health and safety operations at the laboratory level.

In February, 1993, New Jersey's Public Employees OSHA adopted 29 CFR 1910. 1450, Federal OSHA's "Occupational Exposure to Hazardous Chemicals in Laboratories Rule", (also known as the "Lab Standard") *(1)*. The standard provides for the health and safety of employees in the laboratory by generally requiring the components listed below.

1. Development of a written laboratory Chemical Hygiene Plan.
2. Designation of a Chemical Hygiene Officer.
3. Standard operating procedures to be followed when using hazardous chemicals.
4. Consideration for establishment of "designated areas", as well as other protective measures, when working with select carcinogens, reproductive toxins, or particularly hazardous materials.
5. Procedures for the "prior approval" of hazardous laboratory operations.
6. Training of laboratory workers.
7. Medical consultations and examinations for employees with hazardous chemical over exposures.
8. Identification of hazards.

9. Record keeping for any measurements of employee chemical exposures and medical consultations/examinations.

Compliance at New Jersey Institute of Technology (NJIT)

Compliance with the requirements of the published Lab Standard is accomplished by establishing detailed operating procedures and documentation forms included within a larger NJIT Chemical Hygiene Plan *(2)*. Standard Operating Procedures (SOPs) are developed from a set of established criteria that determine necessary control measures *(3)*. SOP implementation is guaranteed by documentation completed by the research investigator, instructor or other responsible party. In addition to SOPs specific equipment must also be assessed on a scheduled basis. This includes fume hoods, personal protective equipment and other laboratory operations. Specific procedures for compliance, and NJIT's review process are discussed below. One key to the NJIT Lab Standard is employee information and training. All laboratory research investigators and laboratory instructors are responsible for providing documented safety training. The intention is to ensure an acceptable level of safety knowledge and workplace practice in the laboratory setting.

Additional Documents

In addition to the written Chemical Hygiene Plan (CHP) a series of additional documents are essential components of the laboratory safety program (Table I). These forms and policies are used to provide additional documentation of compliance for laboratory operations link CHP practice to other safety procedures within the department of Health and Environmental Safety at NJIT.

Table I. Addenda to NJIT's CHP

1. 29 CFR1910.1450 Occupational Exposure to Hazardous Chemicals in Laboratories
2. 29 CFR 1910.1000 Air Contaminants Standard
3. Chemical Spill Handling Procedures
4. Caution Sign Program
5. Accident Report Form
6. Disposal Policy for Laboratory Glass
7. Disposal Policy for Laboratory Chemicals – Pick-up Request Form
8. Disposal Policy for Medical Waste

Scope and Applicability

At NJIT, the Laboratory Standard applies to all employees and students engaged in the "laboratory use" of hazardous chemicals. It applies to research and teaching laboratories which are carrying out small-scale operations (those which can be handled safely and easily by one person) using multiple chemicals and procedures, where the procedures are neither a part of, nor simulate, a production process.

The following personnel are designated with responsibility for implementation of the Laboratory Standard at NJIT. The "*Laboratory Worker*" follows safe work practices, attends required training and is familiar with the laboratory Chemical Hygiene Plan. The "*Laboratory Supervisor*" assures that all employees/students in the lab follow safe work practices, provides necessary hands-on training, develops the laboratory Chemical Hygiene Plan, ensures the Chemical Hygiene Plan is available to all occupants of the lab, and provides "prior approval", when necessary. The "*Chemical Hygiene Officer*" implements the laboratory Chemical Hygiene Plan, provides guidance on safe laboratory procedures, and assists in the annual review and update of the Chemical Hygiene Plan. The "*Department Chair*" designates the Chemical Hygiene Officer (either one for the department, or an individual CHO for each lab), assures department compliance with the standard, appoints department Unit Safety Committee members. The "*University Laboratory Safety and Design Committee*" a subcommittee to the University Health Safety Council) serves as the University Chemical Hygiene Committee, reviews annually the Chemical Hygiene Guide, reviews and approves University policy on laboratory safety.

Standard Operating Procedures

The following standard operating procedures (SOPs) are generic, and apply to most laboratories where chemicals are used. They should be modified, as appropriate, for each specific laboratory. SOPs specific to procedures and operations in each laboratory must be developed and included in each laboratory's CHP. The basic, model information is provided in a CHP template that covers the information listed for each SOP below.

1. *Emergency Procedures* for treating chemical skin contact, behavior in fires and explosions, chemical spill preparedness, hazardous substance identification in emergencies, reporting procedures and behavior during power failures.

2. *General Laboratory Behavior* includes safety rules, additional rules for students, additional rules for supervisors and instructors, rules for custodians, and rules for maintenance workers.

3. *Safety Systems* that provide for personal protective equipment, eye protection, respiratory protection, skin and body protection, hearing protection.

4. *Fire Protection Systems* including explanation of suppression type, alarm configuration, alarm sounds, drills, and outdoor rallying location.

5. *Laboratory Equipment* protocols for items such as fume hoods, glove boxes, eyewashes, safety showers, ground fault circuit interrupters (GFCIs) and spill containment.

6. *General Laboratory Equipment Setup* for preparing the workspace, handling glassware, electricity, vacuum operations, pressure operations, heating, cooling and compressed gases.

7. *Storage Requirements* for maximum allowable flammable storage by class, incompatibility of chemicals in stock, chemical inventory maintenance, Material Safety Data Sheet (MSDS) location and management, and use of refrigerators, flammable cabinets and bench top.

8. *Waste Handling and Disposal Methods* for sink disposal rules, elementary neutralization, "Hazardous Waste" marking requirements, proper chemical name identification, closed container requirements, waste pick-up requests, bulking requirements, and satellite accumulation regulations.

Employee Information and Training

The intent of the Information and Training Program is to inform workers of the physical agents and hazardous chemicals in their laboratory, and the nature of the risks associated with handling these materials. Before working with any of these hazardous materials, lab workers will be informed of the conditions under which the materials may be harmful or may cause injury. They will be trained in the proper control methods (engineering, personal protective equipment, *etc.*) and appropriate procedures necessary to control occupational exposure to hazards in the laboratory. This training is designed to satisfy the requirements of the Public Employees OSHA Occupational Exposure to Hazardous Chemicals in the Laboratories Rule (the Laboratory Standard) and the NJ Worker and Community Right-To-Know (NJ RTK) Act.

The information and training will be provided to University laboratory employees in two separate training sessions. First, a general orientation session will be provided, scheduled, and documented by H&ES. The session covers the topics outlined below, under "General Orientation (Classroom Training)". Second, a "hands-on" training session specific to the employee's work area must be scheduled by the employee's Department and given by their lab supervisor or Chemical Hygiene Officer. This session must cover the items listed below, under "Laboratory Training (Hands On/Specific to Work Area)". H&ES will provide blank attendance forms to document this "hands-on" training session. However, a copy of the completed Department attendance form must be sent to H&ES for compliance documentation.

The SOP for *General Laboratory Behavior* provides training directives for students, supervisors and instructors, custodians, and maintenance workers. The two-part training requirement is an essential component of ensuring that these target groups receive the proper level of safety education.

Training Part I: General Orientation

This is a lecture provided by NJIT's H&ES Department. A regulatory review is provided including an introduction to the CHP and how it applies. An overview of H&ES operations is provided as link between the laboratory and overall campus community. Specific topics include common physical and health hazards, permissible exposure limits (PELs), central files location and use, and chemical exposure prevention.

Training Part II: Laboratory Training

This is a hands-on training specific to each work area provided by laboratory supervisor or principal investigator, as required or needed. CHP availability and SOPs are reviewed. Additional documentation covering emergency procedures, safety equipment, designated PPE and chemical-specific education is delivered. This includes signs and symptoms of exposure and the handling protocols described below. Because this training is an essential part of assessing the need for SOPs and/or special precautions (prior approvals or particularly hazardous chemicals) it is important that the training be conducted by immediate supervisors.

Chemical Hazard Training Content

This section contains descriptions of the general categories of chemical hazard, and the principles of safety associated with each. This section purposefully does not contain advice for handling specific chemicals. Safe work in a chemical laboratory requires very detailed knowledge of the nature, potential, and compatibility's of each substance used; cursory or selective description in this Guide would be misleading and, as a result, unsafe. Additional training is given for specific Hazardous Materials Handling *(4)*.

MSDSs and HSFSs as Cornerstone to Chemical Safety. Anyone planning an experiment or procedure should acquire and review a Material Safety Data Sheet (MSDS) or Hazardous Substance Fact Sheet (HSFS) for each substance, and also for all likely products and byproducts. MSDSs and HSFSs for each chemical present in the laboratory must be available to every laboratory worker within five days of a written request. They are available from H&ES, Specht Building, upon request.

Property-specific Training. The following categories provide a structure for thinking about -- and planning protection against -- common chemical hazards. In actual practice, such hazards do not group themselves in neat categories, but usually occur in combination and/or sequence. The categories and concepts are provided as an aid to awareness, and as encouragement for consistent safe planning and practice.

Flammability. Flammability is one of the most common chemical hazards. The exact degree of hazard, however, depends on the specific substance and the anticipated conditions of use. To handle a flammable substance safely, you must know its flammability characteristics: flash point, upper and lower limits of flammability, and ignition requirements. This information appears on each MSDS or HSFS.

Flash Point. For a liquid, the flash point is the lowest temperature at which the liquid gives off enough vapor to form an ignitable mixture with air and produce a flame when a source of ignition is present. Many common laboratory solvents and chemicals have flash points that are lower than room temperature.

Ignition Temperature. The ignition (or autoignition) temperature of a substance -- solid, liquid, or gas -- is the minimum temperature required to initiate self-sustained combustion. Some ignition temperatures can be quite low (for example, carbon disulfide at 90°C (194°F).

Auto-ignition. Auto-ignition or spontaneous combustion occurs when a substance reaches its ignition temperature without the application of external heat. This characteristic is particularly important to keep in mind in the storage and disposal of chemicals.

Limits of Flammability. Each flammable gas and liquid (as a vapor) has a limited range of flammable concentration in mixtures with air. The lower flammable limit (or lower explosive limit) is the minimum concentration below which a flame is not propagated when an ignition source is present -- such a mixture would be too lean to burn. The upper flammable limit (or upper explosive limit) is the maximum concentration of vapor in air above which a flame is not propagated -- such a mixture is too rich. The flammable range (or explosive range) lies in between the two limits.

Listed measurements of all these characteristics -- flash points, ignition temperatures, limits of flammability -- are derived through tests conducted under uniform and standard conditions that may be very different from actual practice. For example, concentrations of vapor in air in a laboratory are rarely uniform, and point concentrations can be quite high. It is good practice to set maximum allowable concentrations at 20 percent of the listed lower limit of flammability within closed systems. (It is important to note that, generally, this 20 percent limitation is still well above the maximum concentration considered to be safe for health considerations.)

Precautions with Flammable Liquids. Flammable liquids do not burn; their vapors do. For a fire to occur, there must be 1) a concentration of vapor between the lower and upper flammable limits, 2) an oxidizing atmosphere, usually air, and 3) a source of ignition. As it is unlikely that air can be excluded, and unrealistic (given the constant possibility of a spill) to assume that the vapor concentration can be controlled, the primary safety principle for dealing with flammable liquids is strict control of ignition sources.

Ignition sources include electrical equipment, open flames, static electricity, and, in some cases, hot surfaces. Others working in the laboratory should be informed of the presence of flammable substances so that ignition sources can be eliminated. Obviously, it is very important to know which of those sources is capable of igniting a substance you are using. Remember most flammable vapors are heavier than air, and will spread out horizontally for considerable distances until an ignition source is contacted. If possible, flammable liquids should be handled only in areas free of ignition sources. Heating should be limited to water and oil baths, heating mantles, and heating tapes. Static-generated sparks can be sudden ignition sources. When transferring flammable liquids in metal equipment, take care that metal lines and vessels are bonded together and grounded to a common ground. Ventilation is very important. If possible, flammable liquids should only be handled in a chemical fume hood.

Precautions with Flammable Gases. Leakage of compressed or liquefied gases can quickly produce a flammable or explosive atmosphere in the laboratory. This is obviously true where the gases themselves are flammable and under high pressure, but may also be true in the use of non-pressurized liquefied gases. For example, even relatively safe liquefied gases such as liquid air or liquid nitrogen, if kept in open vessels for too long, will generate concentrations of liquid oxygen which can contribute to an explosion. Proper care with compressed gas cylinders and cryogenic setups is essential (see General Laboratory Equipment Setup, Section 5).

Explosiveness. Ignition of flammable vapors or gases can occur with such speed that an explosion results. There are other substances that are explosive in themselves -- in response to heat, mechanical shock, or contact with a catalyst. With some substances, very tiny amounts of impurity are sufficient to begin a reaction that quickly becomes explosive.

Explosives Precautions. Acquire a Material Data Safety Sheet (MSDS) or Hazardous Substance Fact Sheet (HSFS) for each chemical you are using. It is crucial that you know its potential including its compatibilities with other substances. Be alert to any unusual change in the appearance of a reaction mixture. Rapid unexpected temperature rise or fuming are signals for emergency measures such as removing the heat source, quickly applying a cooling bath, or leaving the room. Explosive compounds should be protected from the conditions to which they are sensitive (mechanical shock, heat, light, *etc.*). Check your MSDS/HSFS to see what those conditions are. Such substances should be brought to the laboratory only as required, and only in the smallest quantities absolutely necessary. Reactions involving or producing explosives should be designed on as small a scale as possible, and should be conducted behind a suitable barricade. Special care should be taken that equipment is maintained (for example, that oil is routinely changed in vacuum pumps) and that heating methods used do not cause, or increase the potential for ignition. Other laboratory workers must be notified when an explosive hazard is present, through direct announcement and conspicuous warning signs. Highly exothermic or potentially explosive reactions must never be left unattended.

Explosives Personal Protection. In addition to protection otherwise required in the laboratory, wear face shields, and heavy gloves at all times when handling known explosive substances. Laboratory coats of a flame-resistant material or treatment may help reduce minor injuries from flying glass or flash. When serious explosive hazard is anticipated, shields and barricades will be necessary, along with devices for manipulating equipment at a safer distance long-handled tongs, stopcock turners, mechanical arms, *etc.*). Some experiments at NJIT have required specially designed rooms be constructed for the safety of the researchers. Contact   H&ES if you plan to run an experiment with a significant explosion potential.

Toxicity. Toxicity is the potential of a substance to cause injury by direct chemical action with the body tissues. Whether the effect is acute or chronic, the only way to avoid such injury is to prevent or greatly minimize contact between toxic chemicals and body tissues.

Toxicity Measurement. Dose, or amount of chemical, you are exposed to determines the body's response. In the workplace, there are certain guidelines or regulations which limit your exposure to hazardous substances. These guidelines, which are set by various regulatory or professional organizations are referred to as "workplace exposure limits". A workplace exposure limit is the airborne concentration of a material below which most persons can be exposed for long periods of time without adverse effect. These limits are based on an 8-hour time - weighted-average (TWA) over a working lifetime. Permissible Exposure Limits (PEL) are those set by PEOSHA. Workplace exposure limits may be expressed as Threshold Limit Values (TLV) or Workplace Environmental Exposure Limits (WEEL). Time-Weighted Average (TWA) is the average concentration of a substance integrated over a period of time (*e.g.* a normal 8-hour workday). A Short-Term Exposure Limit (STEL) is the maximum concentration limit for a continuous 15 minute exposure period, provided that the daily TWA is not exceeded. Because workplace exposure limits are generally expressed as average concentrations, excursions above these values are permitted. The exposure levels during such excursions must be below the STEL. However, there are certain levels which must never be exceeded even instantaneously. These are known as the ceiling levels for a TLV, or TLV-C.

All these measurements, though often based on data from animal research, refer to the exposure and resistance of a healthy adult. These levels do not necessarily apply to pregnant women, their unborn fetuses, or adults who are ill or under special stress. In such situations the individual and his/her supervisor or instructor must carefully consider all pertinent information. H&ES can be consulted in such matters.

Acute Toxicity. Acute toxic effects are usually produced by a single large dose, generally well above the TLV, received in a short period of time. The effects are immediate, and may be partially or totally reversible. Acute toxic effects include:

1. *Simple asphyxiation*: the body does not receive enough oxygen (for example, when gaseous nitrogen has displaced the air in a room).
2. *Chemical asphyxiation*: the body is prevented from using oxygen (for example, when carbon monoxide instead of oxygen is absorbed in the blood).
3. *Anesthetic*: causes dizziness, drowsiness, headaches, and coma (for example, by the vapors of many organic solvents).
4. *Neurotoxic*: the brain's control of the nervous system is slowed down or changed (for example, by concentrations of lead and mercury).
5. *Corrosive*: body tissue is directly damaged by reaction with chemicals (for example, by strong acids or bases -- see separate subtopic below).
6. *Allergic*: repeated exposure to a chemical produces sensitization, until there is an allergic reaction at the contact site (usually skin).

Chronic Toxicity. Chronic toxicity refers to adverse or injurious effects that can result from prolonged exposure to a substance, sometimes at dose levels just above the TLV. Damage may not appear for many years -- perhaps generations -- and is often irreversible. As a result, this class of hazard is both very difficult and very important to guard against. The body can filter and process levels of toxicity that might seem surprisingly high, but over extended periods of time, even at the dose very low, the filtering process may fail, and damage may occur. Types of chronic toxic effects include:

1. *Carcinogenicity*: produces cancer (for example, asbestos and vinyl chloride are known to produce cancer in humans).
2. *Mutagenicity*: alters cell genes; subsequent generations show genetic damage.
3. *Teratogenicity*: harms developing fetus.
4. *Reproductive toxicity*: interferes with the reproductive system in men or women.
5. *Specific organ toxicity*: damages specific organs (for example, carbon tetrachloride can cause liver damage).

Precautions with Toxins. The precautions to take against contact with toxic substances are repeated many times throughout this Guide. With chemicals of low acute toxicity, it may be tempting to be less rigorous; yet it is precisely those chemicals which most require continual caution -- an unvarying habit of safety.

You must protect your body against all forms of chemical contact: absorption, inhalation, ingestion, and injection. Never eat, drink or smoke in the laboratory; wear the appropriate protective gear, and always remove it before you leave the laboratory. Make sure you carefully wash your hands before leaving the laboratory. Remember that the chemicals you bring home on your clothes will have a more powerful effect on growing children and elderly people than on most adults. In order to know what level of personal protection will be adequate, keep up to date on recent tests for substances you are using. MSDSs are updated regularly, and you should consult the most recent data each time you begin a new procedure. The best precaution is to treat all chemicals as toxic.

Corrosives. Corrosivity is a form of acute toxicity sufficiently common and hazardous to merit separate discussion. Corrosive chemicals include strong acids, strong bases, oxidizing agents, and dehydrating agents. When they come in contact with skin, eyes, or, through inhalation, the surface tissues of the respiratory tract, they react with the tissues they touch and cause local injury.

Liquid Corrosives. A liquid corrosive will act on the skin rapidly or slowly depending on concentration and length of contact. These chemicals react directly with the skin: dissolving or abstracting from it some essential components; denaturing the proteins of the skin; or disrupting the skin cells. Mineral acids, organic acids, and bases are among the typical liquid corrosives. When handling liquid corrosives, contact with them must be scrupulously avoided. Wear goggles, rubber or suitable synthetic gloves, and a face shield. A rubber or synthetic apron and rubber boots may also be necessary. Since many liquid corrosives also release irritating vapors, procedures using these materials should be performed in a fume hood.

Solid Corrosives. Solid corrosives interact with the skin or other surfaces when dissolved by the moisture there. Damage then occurs both from the corrosive action and from the heat of solution. Because they are solid, these chemicals are relatively easy to remove; but because they may not react immediately and may not be painful at first (as with the caustic alkalis), they may cause much damage before being detected. Solid corrosives are most commonly dangerous in a finely divided state. Dust control and good exhaust ventilation are essential, as well as goggles, gloves, and other protective clothing. In case of chemical contact, much care must be taken during the emergency shower irrigation to remove all particles of solid matter that might be lodged in the skin or clothes.

Gaseous Corrosives. Gaseous corrosives pose the most serious health hazard of all corrosives because of possible damage to the lungs, including spasm, edema, pneumonia, and even death. Different corrosive gases affect different parts of the lung (for example, ammonia affects the upper respiratory tract, while phosgene affects the lung, causing pulmonary edema), but all are to be avoided. It is thus crucial that corrosive gases not be inhaled. Careful design and the use of fume hoods is essential. Skin and eyes must also be protected, as gases contact all exposed parts of the body.

Impurities and Combinations. MSDSs contain information on pure chemicals, known mixtures, and proprietary materials. Unfortunately there are no such sheets for other materials found in the laboratory, including solutions, mixtures of unknown or uncertain composition, and byproducts of reactions, all common in the laboratory. Impurities, synergistic effects, formation of unexpected products and byproducts, insufficiently clean equipment, and the combination of vapors from your experiment with that of your neighbor's can all produce sudden and unanticipated hazards.

There is no absolute protection against all contingencies, but it helps to wear protective gear, to clean equipment scrupulously, to be aware of experiments in progress in nearby areas, and to be completely familiar with emergency procedures.

Training is the Cornerstone to Additional Documentation

The training information provided above serves a a template for written programs, SOPs, specific protocol and documentation. The training process points toward needed assessments and further prior approvals as shown below. Specific control procedures may be identified during training as a result of the dessemination of basic chemical safety knowledge and the legal requirements of the CHP.

Review Forms Indicate Training and Generate SOPs

In general a review form is maintained that provides a training summary. A review form can serve as an inspection guide, SOP template and written CHP template. A form generally includes, but is not limited to the basic content provide in Table II. The basic training explained above will comply with most SOPs and help ensure a high level of safe behavior. The purpose of a review form is to point to further assessments needed for prior approvals and/or particularly hazardous chemicals as discussed below.

Table II. Basic Review Form Contents

1. Background of specific chemicals to be used.
2. Personnel involved with the project.
3. Experimental procedures summary.
4. Recognized control procedures utilized.
5. Decontamination and disposal protocols.
6. Emergency Procedures.
7. Monitoring procedures.

Additional Operations Requiring Prior Approval

In general, prior approval must be obtained when a laboratory procedure presents a significant risk of injury, illness, or exposure to hazardous substances. The risk is considered significant when there are very large quantities of particularly hazardous substances involved or the experimental procedures exacerbate the potential for a hazardous condition. These conditions must be applied on a case-by-case basis. The process of documented assessment ensures that all relevant safety parameters are met, including training. The documentation used is summarized in Table III. The information is completed and reviewed by the Principal Investigator and the Unit Safety Committee.

TableIII. Safety Assessment for Specific Operations

Safety Approval and Documentation Form

Name of operation, date, location, Principal Investigator, Department.

- List all individuals who have been trained in this procedure.
- General_Provide a brief description of the activity which will be carried out.
- Identify and addressed all hazards associated with materials, equipment and procedures.
- Summarize the hazards which may be encountered, including: toxicity, flammability, pressure, vacuum, temperature extremes, noise, explosivity, etc.
- Identify the written procedures for what you are doing.
- Indicate location of current copies of MSDSs.
- Indicate that all MSDS information has reviewed by relevant users.
- Have all individuals been trained and do they understand the written procedures?
- Has the potential for emergency situations been addressed (*e.g.* runaway reaction, loss of temperature control, etc?)
- Are shut-offs for bottled gases or other critical valves/shut-offs located where they can readily and safely be reached and closed?
- Are specific emergency shut-down instructions posted and visible?
- Post emergency shut-down procedures for all overnight and unattended operations. Ensure that there is a Caution Sign posted on the laboratory door.
- Is appropriate protective equipment (*e.g.* gloves, goggles, face shields, lab coats, *etc.*) available and being used?
- Are all individuals familiar with what to do in the event of accidental contact (*e.g.* inhalation, ingestion, skin contact)?
- Are all individuals familiar with what to do in the event of a spill or emergency?
- Are you using the least hazardous materials and minimum practical quantities for your needs?
- Is appropriate safety equipment available and in working order (*e.g.* fume hoods, glove boxes *etc.*)?
- If any question above is answered with a "NO", please explain, below.

Provisions for Particularly Hazardous Chemicals

Particularly hazardous chemicals include carcinogens, reproductive toxins, acutely toxic substances, and chemicals of unknown toxicity. Specific CHPs must be developed and documented for these situations using the information provided above as a minimal template. All protocols, training, PPE, inspections and other CHP specifics must be fully developed and documented. The control procedures for particularly hazardous chemicals are site-specific.

Implementing Identified Control Procedures

Additional assessment tools are used by the NJIT H&ES Department. Whenever possible, MSDSs for chemicals used in the laboratory will be reviewed prior to the use of a chemical. These data, along with information on the conditions under which the chemical is to be used, will generally be used to determine the degree of protection required. In certain circumstances, H&ES will conduct exposure monitoring to determine adequacy of controls and to determine if additional control measures are necessary. Published occupational exposure limits are reviewed and observed. Exposure potential is determined by reviewing experimental procedures. Exposure potential is generally increased with increased temperature or pressure, when working with open rather than closed systems, during transfer of materials, during the use of hazardous substances with live animals. Exposure can occur via inhalation, skin contact (with liquid, solid or vapor) or through accidental ingestion. Generally, greater exposure potential requires higher levels of protection. Once the required the degree of control is determined, control measures will be implemented in the following 1,2,3 hierarchy.

Priority 1: Engineering Controls. Engineering controls reduce an exposure at its source. Engineering controls are the method of choice for reducing exposures and will be used whenever possible/practicable. Examples of some engineering controls include:

1) Substitution of hazardous materials or operations with those which are less hazardous
2) Use of Laboratory fume hoods
3) Use of glove boxes or other enclosures
4) Use of local exhaust ventilation (e.g. "elephant trunks", slotted exhaust hoods, and canopy hoods).

Priority 2: Administrative Controls. Administrative controls are work practices which are designed to control exposures. Administrative controls will be used in conjunction with engineering controls or when engineering controls are impractical or not feasible. Examples of administrative controls include:

1) Limiting time of exposure to maintain levels below acceptable exposure limits.
2) Utilizing good housekeeping procedures to reduce exposures.

Priority 3: Personal Protective Equipment. Personal protective equipment does not reduce the source of exposure, but rather protects the individual. Personal protective equipment will be used in addition to engineering controls, while engineering controls are being installed or when engineering controls are impractical or not feasible. Some examples of personal protective equipment include:

1. Respirators -- This includes dust masks, as well as other types of respiratory protective equipment). Because all respirator users must participate in NJIT respiratory protection program, you must call H&ES if you think you have a need for respiratory protection.

2. Gloves, aprons, boots, and other skin protection, goggles and face shields. Check the glove manufacturers compatibility chart to ensure using the proper glove material for the solvent being used.

Training in Practice: Preparing for Laboratory Work

Before beginning any laboratory work, a plan should be made describing: goals; chemicals and equipment needed; and the sequence of steps to be followed, including safety measures. This plan also serves as the template for laboratory-specific training. Because the Laboratory Standard is "performance" oriented, it is important that these laboratory work procedures are implemented in daily behavior.

General Chemical Usage. Full descriptions of chemicals used in the laboratory can be found on Material Safety Data Sheets (MSDSs) or Hazardous Substance Fact Sheets (HSFSs), which contain information on physical characteristics, hazards, disposal, and routine and emergency precautions. There is a sheet for virtually every chemical marketed, available from chemical suppliers, University RTK Central Files (available in Police Dept.), H&ES, and a number of computer based information systems. HSFSs are available from the NJ Department of Health for each of the substances regulated by the New Jersey Worker and Community Right-to-Know law. The Right-to-Know law requires, among other things, that persons who may be exposed to chemicals be trained in general and specific chemical hazards and chemical safety. MSDSs and HSFSs should be used as part of this training. An MSDS or HSFS should be acquired for every chemical used and should be kept on file for reference. The information on the MSDS or HSFS should be given to every laboratory worker who will be handling the chemical in question. Design your procedure to use the least hazardous chemicals and the minimum possible quantity of each chemical that will still allow meaningful results. Using smaller quantities of chemicals means that less can be spilled or volatilized, and that less must be treated and/or disposed as hazardous waste.

Chemical Equipment and Instrumentation. Specific information must be obtained about any equipment to be used. Most equipment is sold with this information, ranging from one page instruction sheets to complete books. This information must be read thoroughly and followed exactly for safest use of the equipment. When used equipment is sold or donated to the University, recipients must obtain operating instructions if at all possible.

CHP Written Procedures for SOPs. Developing a protocol is basic to the experimental process, and should result in a written set of procedures. Writing the procedures allows the researcher or instructor to go through the experiment in the planning stage, and identify areas where special precautions may be necessary. The written protocol will provide workers with step-by-step instructions, minimizing the chance of errors. A good written protocol will allow for modifications and will include safety precautions (*e.g.*, "wear splash goggles," "pour acid into water," "perform this operation in fume hood"). Written procedures should also include MSDSs or HSFSs for all chemicals used in the experiment. In addition, a laboratory notebook should be kept during the procedure, documenting each action and its result. In the event of an accident, a set of written procedures and laboratory notebook may indicate what went wrong, and possibly why.

Setting up the Laboratory. Just before beginning the work, review the written procedures, following the expected sequence of the experiment. Review the materials to be used as to their degree and nature of hazard, including flammability, volatility, reactivity, *etc.* All equipment and supplies should be in place before actual work begins, including proper protective equipment (*e.g.*, hoods, glove boxes, gloves, aprons, safety goggles, shields, and lab coats). The work area should be uncluttered and orderly. Where areas of possible contamination and exposure might exist, take precautionary measures, such as lining the work surface with absorbent paper. Also, have on hand all the necessary equipment to deal with a spill or accident (more absorbent paper, spill-control kits, *etc.*)

Handling Chemicals Guidelines. Following are guidelines and principles for safety in the direct manipulation of chemicals -- holding, pouring, mixing, transporting, storing, and so on. The list of situations covered is far from exhaustive; emphasis is instead on the most common ways in which chemicals are handled in the laboratory. Safety precautions for use of laboratory equipment can be found in Safety Systems, Section 3, and General Laboratory Equipment Setup, Section 5.

Responding to Personal Contacts with Chemicals. The primary safety goal in handling chemicals is to prevent the chemicals from entering your body. It cannot be said too often that protective gear must be worn at all times, and precautions for avoiding personal contact with the chemicals must always be in mind.

1. Avoid direct contact of any chemical to the hands, face, and clothing.
2. Be aware of what you touch; be careful not to touch gloves to your face, for example.
3. After any skin contact, and always before you leave the laboratory, wash face, hands, and arms.

4. Leave all equipment in the laboratory.
5. Never taste chemicals or sniff from chemical containers.
6. Never eat, drink, smoke, or apply cosmetics in the laboratory.
7. Dispense and handle hazardous materials only in areas where there is adequate ventilation.
8. If you believe that significant ingestion, inhalation, injection, or skin contact has occurred, call the emergency number for your campus and f ollow the Emergency Procedures given in the CHP.

Handling Containers. Clearly label all chemical containers. The Laboratory Standard requires that labels on incoming chemical containers not be removed or defaced. Do not use any substance from an unlabeled or improperly labeled container. Printed labels that have been partly obliterated or scratched over, or crudely labeled by hand, should be relabeled properly. Unlabeled chemical containers are a violation of the NJ Right-to-Know Act and should be disposed of promptly and properly. Carefully read the label before removing a chemical from its container. Read it again as you promptly recap the container and return it to its proper place. Names of distinctly different substances are sometimes nearly alike; mistakes are easy to make and can be disastrous. When picking up a bottle, first check the label for discoloration, and if it is clean, grasp it by the label. Spilled chemical will show up on the label better than on the glass; holding the container by the label will protect you from prior spills, and protect the label from present ones. After use, wipe the bottle clean. If a stopper or lid is stuck, use extreme caution in opening the bottle. Friction caused by removing tops may cause explosions with some substances (such as hydroperoxides formed from ethers or picric acid contaminated with heavy metals). Support beakers by holding them around the side with one hand. If the beaker is 500 ml or larger, support it from the bottom with the other hand; also, consider using a heavy-duty beaker and slowly placing it on the clean surface of the bench. If the beaker is hot, use beaker forceps or tongs, and place the beaker on a heat-resistant pad. Grasp flasks by the center neck, never by a side arm. If the flask is round bottomed, it should rest on a proper sized cork ring when it is not clamped as part of a reaction or distillation assembly. Large flasks (greater than 1 liter) must be supported at the base during use. Never look down the opening of a vessel, in case of unforeseen volatility or reaction.

Pouring. Do not pour toward yourself when adding liquids or powders. Stoppers too small to stand upside down on the bench should be held at the base and outward between two fingers of the pouring hand. Use a funnel if the opening being poured into is small. If a solid material will not pour out, be careful when inserting anything into the bottle to assist removal. Students should seek advice from instructors before proceeding. Always add a reagent slowly; do not "dump" it in. Observe what takes place when the first small amount is added and wait a few moments before adding more. When combining solutions, always pour the more concentrated solution into the less concentrated solution or water. Stir to avoid violent reactions and splattering. The more concentrated solution is usually heavier and any heat evolved will be better distributed. This procedure is particularly applicable in preparing dilute acid solutions.

40

Be sure to wear goggles and use the hood when diluting solutions. Make sure the stopcock is closed and has been freshly lubricated before pouring a liquid into an addition or separatory funnel. Use a stirring rod to direct the flow of the liquid being poured. Keep a beaker under the funnel in the event the stopcock opens unexpectedly. Wear an apron and gloves, in addition to goggles, whenever pouring bromine, hydrofluoric acid, or other very corrosive chemicals, to avoid painful chemical burns.

Pipetting. Never pipette by mouth. Use an aspirator bulb, or another mechanical pipetting device. Constantly watch the tip of the pipette and do not allow it to draw air.

Summary of NJITs CHP

The CHP template used at NJIT is designed to produce necessary SOPs, describe overall campus operations relating to laboratories and ensure a proper, high level of training and education for all workers effected by laboratory operations. Documentation of training and procedures is a key element of compliance, but also serves to maintain a high level of chemical safety communication on the NJIT campus.

References

1. *Occupational Exposure to Hazardous Chemicals in the Laboratory;* 29 CFR 1910.1450; Occupational Health and Safety Administration, 1990.

2. *The New Jersey Institute of Technology Chemical Hygiene Plan;* Van Houten, Norman J.; NJIT; Newark, NJ. 1996.

3. *Laboratory Safety Standards for Industry/Research/Academe*; Van Houten, Norman J.; Scitech Publishers Inc., Matawan, NJ. 1990.

4. *NJIT Hazmat rsponse Team HAZWOPER Study Guide.* Department of Environmental Health and Safety; NJIT. 1995.

Chapter 4

Laboratory Waste Management and Waste Determinations

Russell W. Phifer

Environmental Assets, Inc., 909 Old Fern Hill Road, Suite 1,
West Chester, PA 19380

Laboratories generate considerably less than one tenth of one percent of all hazardous wastes. Individual academic laboratories, in particular, produce minuscule amounts of these regulated wastes. So why is laboratory waste such a problem? Simply put, no other class of waste generator produces anything close to the number of different chemical wastes of a laboratory. Because of the regulatory structure, this creates enormous disposal problems for labs, and an extremely high per unit cost for disposal. In addition, the requirement to characterize wastes prior to disposal presents a recordkeeping nightmare unlike that of any industrial facility. Considering all these problems, then, why should hazardous laboratory wastes be regulated? The obvious answer is that laboratory waste can cause problems to the environment and to human health if not handled properly. The regulations are based on the assumption, for example, that chemicals in landfills will eventually leach out and impact groundwater supplies. It is for this reason that hazardous wastes can no longer be landfilled.

What is regulated as a hazardous waste? There have been only minor changes since the passage of the Resource Conservation and Recovery Act (RCRA) in 1976; this law was first implemented in 1980. The law states that any discarded chemical or chemical-byproduct that either meets a specific characteristic or is included on one of four lists is regulated as a hazardous waste. The four possible characteristics of a hazardous waste are shown below.

Ignitability. Liquid ignitable hazardous waste has a flash point less than 140 ^{0}F (60 ^{0}C) or has some other characteristic that has the potential to cause fire. Solid materials are regulated as ignitable if they are capable of causing fire through friction, absorption of moisture or spontaneous chemical changes. Generally, any solid material that burns vigorously and persistently is considered ignitable. Flammable gases and oxidizers also meet the characteristic of ignitability.

Examples of ignitable chemical waste that might be found in a laboratory:

Flammable liquids	Organic solvents such as acetone, alcohols, toluene, xylene
Flammable gases	Hydrogen, silane, butane
Oxidizers	Nitrate salts, peroxides

These materials are potentially dangerous to human health and the environment because they make excellent fuel for a fire. Regulation 40 CFR 261.21 describes the characteristic of ignitability in detail. Ignitable waste is identified on manifests and other reports with the identification number of D001. Some ignitable waste organic solvents are also regulated by EPA because they are listed separately in 40 CFR 261 in that Part's F, U or P lists. Waste identification numbers F001 through F005 describe certain spent (waste) solvents. Wastes that are hazardous due to listing in one or more category must have all EPA waste numbers provided on shipping documents.

Corrosivity. Liquids are regulated as corrosive hazardous waste if they have a pH less than or equal to 2, or greater than or equal to 12.5. Under current regulations, a solid material cannot be regulated as a corrosive waste. Corrosives are either acidic or basic; an acid is a substance that can give up a proton, while a base is a substance that can combine with a proton to form a new compound. Ironically, the EPA initially intended to regulate the acid range up to a pH of 2.5; this had to be changed when they realized that popular soft drinks such as Coca Cola and Pepsi contain Phosphoric acid, and have a pH of about 2.3!

Examples of common corrosives found in laboratories include:
Sulfuric acid
Hydrochloric acid
Phosphoric acid
Ammonium hydroxide
Sodium Hydroxide

Why should these materials be regulated as hazardous wastes? What is their potential impact on human health and the environment? One primary concern from corrosives is that if they are thrown in with regular trash, workers involved in collection may come in contact with materials that attack clothing and skin. Indeed, there are numerous documented cases of trash collectors being hospitalized from contact with corrosives. In addition, strong acids can accelerate the breakdown and release of toxic heavy metals from many consumer products that might be disposed of as routine municipal waste. One example is metal furniture plated with chrome.

D002 is the identification number for wastes that are hazardous due to the corrosivity characteristic.

Reactivity. Reactive hazardous waste includes chemicals that are unstable, readily undergo a violent change, react violently with water, or are capable of detonation or an explosive reaction if subjected to a strong initiating source. A cyanide or sulfide bearing waste is also considered reactive, as are other wastes that have the potential to generates toxic gases, vapors, or fumes.

Examples of reactive or potentially reactive waste include:

Reactive	Sodium, potassium and other alkali metals
Potentially explosive	Dry picric acid; ether that contains peroxides
Toxic gas source	Cyanide or sulfide solutions

40CFR 261.23 describes this characteristic in detail. Reactive waste has the identification number of D003.

Toxicity. EPA regulates toxic chemical waste in two ways. First, it defines a characteristic of leachate toxicity. If disposed of in a landfill, waste with this characteristic would create leachate (liquids that drain through or from waste) containing toxic metals, pesticides or certain other chemicals, thus posing a threat to groundwater. The complete name of this characteristic is TCLP (Toxicity Characteristic Leaching Procedure) Toxicity. The remaining D identification numbers in 40 CFR 261.24 are used for waste exhibiting this characteristic. This characteristic brings many wastes into regulation, especially those wastes containing only trace amounts of contaminants -- and wastes you may not suspect of being hazardous.

In addition to the hazardous waste characteristics, there are also four lists of chemicals (U, P, K, and F) that are regulated as hazardous. These are described below.

The P and U lists include commercial chemical products and waste from the cleanup of spills from these materials. P and U listed waste does not include materials contaminated with P and U list chemicals, such as chemically contaminated gloves, etc; this type of waste is not regulated by EPA unless it meets a characteristic definition.

P list chemicals are termed acute hazardous waste. It is important to identify acute hazardous waste because generation of more than 1 kg/month of P chemicals will require the laboratory to comply with the most stringent generator requirements (see below). The list of acute hazardous waste includes allyl alcohol (P005), arsenic pentoxide (P011), carbon disulfide (P022), cyanides including soluble cyanide salts (P030), nicotine and salts (P075), osmium tetroxide (P087) and parathion (P089).

U list chemicals are generally more common and less toxic than P list chemicals, but both refer only to discarded commercial chemical products. Examples of U listed chemicals include ethyl ether(U117), formaldehyde(U122) , phenol (U188), and Toluene (U220).

The F list is for wastes from non-specific sources such as cleaning or degreasing. Many laboratories routinely generate F listed wastes as spent solvents; the F list includes such solvents as xylene, acetone, methanol, and methylene chloride.

The K list is for wastes from specific sources, such as wastewater treatment sludge from the production of pigments, API separator sludge from the petroleum refining industry, and spent carbon from the treatment of wastewater containing explosives.

Be aware that EPA regulations were generally meant for industry and do not cover some hazardous chemical wastes that are generated in laboratories. However, many unregulated wastes deserve special precautions.

Regulatory requirements

There are numerous regulatory requirements associated with the handling of hazardous wastes. These include waste determinations, recordkeeping, reporting, storage, accumulation, and labeling. These responsibilities are typically performed (or supervised) by someone who has received RCRA training, as mandated by the federal law and state regulation. This is important from a safety perspective as well as to meet regulatory requirements. Most states have received authority from the US Environmental Protection Agency (EPA) to issue their own permits and regulations, though they must be at least as restrictive as the federal model. The primary concern of RCRA is the cradle to grave tracking of hazardous wastes - from the time they are first generated until the time of their ultimate disposal.

Waste Determinations
The first step in any effort to manage wastes is a determination of whether or not each specific waste "stream" is hazardous by RCRA definition, and if so, designation of hazard class. Waste determinations can be made either from knowledge of the waste or through analytical testing. Obviously, the easier and least expensive way is through knowledge of the waste - what went into the waste container, and the specific volumes. If accurate records are not available, then testing is necessary to determine if the waste either meets a characteristic or if a listed compound is included. Maintaining accurate inventories and properly labeling all waste containers can save time and money when the time comes to

prepare for disposal. All students and laboratory workers who generate wastes should consider this their responsibility.

Recordkeeping
It is important to maintain documentation of hazardous waste management activities for a variety of reasons. Many records are required to be kept for a specific period of time; manifests of shipments, for example, must be kept for a minimum of three years from the shipment date. In reality, however, records should be kept permanently, since it may be necessary to document what was included on a specific shipment if the generator is subject to a CERCLA (Superfund) action.

Reporting
Generators are required to complete biennial reports to the US EPA. These reports, which are due by March 1 of even numbered years, document generator information, describe waste minimization efforts, and show all shipments of hazardous waste shipped off-site to storage, treatment and disposal facilities. Individual states may also have reporting requirements.

Storage
Large generators are required to meet specific standards for the on-site storage of hazardous waste. Included in the requirements are provisions for containment, management of containers, storage area communication, and emergency planning.

Accumulation
Perhaps the least understood requirements of RCRA are those that apply to the accumulation of hazardous wastes. As previously noted, generators cannot store on-site longer than set time limits which are based on monthly rates of generation. There are, however, provisions for the storage of waste at the point of generation. This is limited to 55 gallons of waste per instance, and wastes must be stored in the immediate work area.

Labeling
Hazardous wastes must be properly labeled during storage as well as for transportation. Labels must be clearly visible and legible, and contain such information as hazard class, generator name, and accumulation start date. For transportation, the wastes must also show EPA and US Department of Transportation (DOT) reference numbers.

How RCRA Applies to Laboratories
In many ways, applying hazardous waste regulations to laboratories is a lot like trying to put a square peg in a round hole. RCRA and its resulting regulations were designed for industry, not laboratories. While the requirements still apply, many laboratory operations can take advantage of flaws in the regulations. For example, many written laboratory procedures, both in academic and R&D

laboratories, can be changed to add a step that neutralizes reaction by-products or renders them less hazardous. There also are certain exemptions related to wastewater that might apply to your laboratory. RCRA exempts laboratory wastewaters that would be hazardous due to corrosivity or toxicity as long as the volume of laboratory wastewater does not exceed 1% of the entire facility's wastewater. However, local sewer authority regulations still apply.

Laboratory treatment of hazardous wastes
While treatment of hazardous wastes cannot generally be performed without a permit, adding a step to a written procedure in an academic setting not only minimizes wastes generated, it is an effective teaching tool that a student can take with him to a laboratory setting in industry. By getting students thinking about the by-products and wastes generated during laboratory procedures, they are encouraged to consider all the implications of lab processes, including waste disposal. One example is precipitating a metal that would be regulated as hazardous, such as Arsenic or Chromium. Trivalent chromium can be treated with lime or caustic to raise thepH to above 8 to precipitate chromic and other metal hydroxides that would be more stable. This also reduces the volume of waste. It is important to note that laboratory waste treatment of this type is not allowed without a permit; it is for this reason that such treatment must be part of the written procedure, prior to declaring the by-product of a procedure a waste.

Innovative treatment techniques can also be investigated (under specified conditions) as an exempt "treatability study". Since waste treatment and disposal are both expensive (frequently costing more than the original cost of the chemicals), the development of new treatment processes is a valuable and necessary part of laboratory research.

It should be noted that there are several treatment processes that are allowed and encouraged under RCRA. These include acid-base neutralization and treatment in an accumulation container. Mixing together an acid and a base generates a salt and water, which can be easily disposed of down the drain. Obviously, proper laboratory techniques should be utilized to eliminate violent reactions; materials should be added together slowly, and proper skin protection is necessary in the event of splashing.

While neutralizing an acid solution with a base (or vice versa) is a relatively simple process when done properly, treating wastes in a container used for accumulation can be considerably more risky. Since RCRA regulations require that waste containers be kept closed except when wastes are being added, treatment must not be exothermic reactions. In addition, it is necessary to have absolute knowledge of all wastes that have been added to the container. Such treatment should not be attempted without considerable planning and full anticipation of the end result. It should be noted that safety is a crucial part of the waste management process; safety considerations include the selection of appropriate personal protection equipment, compatible containers, and verifying the compatibility of any chemicals that are mixed.

Laboratory exemptions

Aside from the previously mentioned provisions for treatability studies (which require notification to the state agency and the EPA Regional Administrator), one of the most interesting exemptions is for wastewater discharges to a publicly owned treatment works (POTW). Laboratory wastewaters are exempt from RCRA (for corrosive and toxic discharges only), as long as the total volume of wastewater does not exceed 1% of the entire facility's discharges. While local sewer authority regulations still apply, written permission may be obtained from many such authorities for "laboratory scale" discharges of hazardous wastes. Other limitations may apply, so careful consideration should be given before pouring hazardous wastes down the drain. For example, discharges of greater than 15 kilograms per month may require continuous discharge notifications under CERCLA.

Laboratory waste minimization

There are many situations where laboratory wastes can be effectively minimized. The best starting point is with a review of purchasing and inventory control procedures, since a large percentage of laboratory wastes consists of discarded chemical reagents. Users should consider the economics of chemical use, which includes storage and disposal, as well as the initial purchase cost.

Waste minimization efforts from a laboratory prospective should have a strong focus on source reduction. Source reduction consists of any and all means to reduce the volume of chemicals used, which results in less waste being generated. Aside from maintaining an accurate inventory and establishing effective inventory control measures, laboratory procedures should be reviewed to determine if smaller quantities of reagents can be used effectively. Microscale equipment is readily available from a number of suppliers; while working with smaller quantities may result in the need to make modifications to procedures, it also forces students (and lab employees) to pay more attention to quality control measures. Other source reduction methods include product substitution (replacing a hazardous chemical in a procedure with one that is less hazardous) and process changes (reducing the number of steps in a procedure, or working with different equipment). Many laboratories have successfully integrated computer simulations of experiments into their programs; this effectively eliminates the use of chemical reagents.

Laboratories should also consider reclamation / recycling of waste solvents. By redistilling and reusing laboratory solvents, the purchase cost of new material can also be saved. If the quality of reclaimed solvent is not suitable for a specific procedure, there may be other uses for the solvent. Examples include using reclaimed solvent as a fuel supplement in industrial boilers, using for cleaning purposes, or as a thinner for paints or inks by a graphics department.

Off-site disposal of laboratory wastes

Preparing for of-site disposal of laboratory wastes is a multi-step process that should be part of every facility's planning. The process includes preparing wastes for shipment, selecting disposal facilities and vendors, and contracting for transportation and disposal. While much of this work should be performed by individuals with RCRA training (to meet regulatory requirements), all students and workers should be aware of the steps involved in off-site disposal.

Selecting disposal facilities and vendors
The decision making process for selecting disposal facilities and vendors includes a review of potential costs and liabilities associated with different disposition technologies and the contractors that handle transportation and disposal. While cost should be a major consideration, long-term liability for proper disposal is even more important. Since generators are responsible for wastes from "cradle to grave", it is crucial that disposal be permanent - that interim storage of wastes off-site be minimized. Disposal technologies should ideally result in the complete destruction of wastes. While incineration most clearly meets this goal, chemical treatment can also have the same result, since wastes lose their "identity" when treated and rendered non-hazardous. Generators should review the steps involved in off-site disposal, including any interim storage, to verify that future liability will be minimized.

Preparing wastes for off-site shipment
The more a generator can do to properly prepare wastes for shipment, the lower the cost for transportation and disposal. For instance, compatible wastes should be bulked to reduce the number of containers to be shipped. From a transportation and disposal perspective, the container size is more important than the actual volume of waste. A half-empty drum takes up as much space as a full drum, and there are no disposal discounts for partially full drums. Proper segregation and labeling can also help reduce costs and future liability; should a container be improperly characterized or labeled, the generator could have future criminal or civil liability if an accident should result in harm to human health or the environment.

Chapter 5

Motivating Laboratory Instructors at a Major University To Formally Incorporate Safety Education into Their Curricula

R. B. Schwartz

Director of Environmental Health and Safety, Oakland University,
Rochester, MI 48309–4401

Developing a genuine safety culture in the laboratories of an educational institution requires the formalized safety "training" of faculty and staff, in addition to (and perhaps moreover) the academic students in those laboratories. This paper presents a case study at Oakland University (OU), a mid-western suburban residential college (annual FTE = 9,835), wherein the university's Office of Environmental Health and Safety (EH&S), in conjunction with its Laboratory Safety Committee, LSC, (Chaired by the Director of EH&S), conceived and executed a **Laboratory Student Safety Program** (LSSP). The overall intent of the LSSP was to bridge the gap between a) the "chemical hygiene training" required by 29 CFR 1910.1450, (provided by EH&S to the university's science and research faculty), and b) the less "homogenous" safety education being presented by these instructors to their lab students.

It should be noted that, while the primary *content* of OU's LSSP (and the case study presented herein) focuses on Oakland University's chemical laboratories, the program manual reminds its readers that the *intent* of the document should be universally applicable to any "hands-on" instructional setting (thereafter referred to as "laboratories") in which biological, electrical, mechanical, physical, or chemical hazards are presented. Accordingly, the Program was designed for "laboratory" instructors in the departments of Biological Sciences, Chemistry, Engineering (Mechanical, Electrical/Systems, Computer), Health Sciences and Physics.

Background

Oakland University established its Office of Environmental Health and Safety (EH&S) in February 1990. It was not until 1991 however, with the advent of the Bloodborne Pathogens and Laboratory Right-to-Know Standards, that EH&S began to interact directly with faculty regarding EH&S issues (and the regulatory requirements thereof). The relationship developed slowly, and not without faculty resistance. The Laboratory Safety Committee, LSC (chaired by the Director of EH&S), was established as a non-governing (advisory) body by OU's President in 1994, in response to a laboratory fire. LSC representation/distribution appears in Table I.

Table I.   Oakland University Laboratory Safety Committee Representatives

Department	# of Representatives	Distribution
Biological Sciences	3	1 Chairperson (Faculty) 1 Laboratory Manager 1 Stockroom Manager
Chemistry	3	1 Faculty 1 Laboratory Manager 1 Stockroom Manager
Clinical Research Laboratory	1	1 Manager
School of Engineering	1	1 Laboratory Manager
Office of EH&S	3	1 Director (& LSC Chair) 1 Indus. Hygiene Coor 1 EH&S Intern
Eye Research Institute	2	1 Principal Investigator 1 Asst to Director
School of Health Sciences	1	1 Faculty
Physics	1	1 Laboratory Manager

Stage I: Preparation of the LSSP (July - September 1994)

Introducing the LSSP *Concept* to the Laboratory Safety Committee (LSC). The first goal of OU's EH&S Director/Chair of the LSC was to introduce the *need* for a "Laboratory Student Safety Program" (LSSP) at Oakland University. Since the LSC members were selected based largely on their historical abilities to favorably influence colleagues, unanimous LSC support of the *concept* was essential to the success of the program. Initial LSC resistance was not surprising; it is summarized in Table II.

Table II.   Stage I - LSC Resistance to and Acceptance of the LSSP Concept

Laboratory Safety Committee Points of Resistance to LSSP Concept

◆ General impression that laboratory instructors DO maintain *safe* laboratories (coupled with a negligible history of laboratory injuries/illnesses)

◆ Assertion that a vast majority of (particularly undergraduate) laboratories do not perform particularly hazardous operations

◆ Overwhelming contention that faculty would never permit an administrative body (e.g., the Office of EH&S and/or the LSC) to mandate *any* components of their curricula

Challenges to the LSC Points of Resistance

◆ History of communications from lab students to EH&S indicating a question-able presence of consistent guidance regarding chemical and waste handling

◆ EH&S assertion that injury and illness reports are statistically very *poor* indicators of safe conditions/acts

◆ Distinction between maintaining safe conditions (particularly in facilities where hazard potential is inherently low) and conscientiously *educating* students regarding the *design* of safe protocols and conditions, (the latter approach constituting the students' only potential for obtaining safety tools, which can then be used in future, perhaps more hazardous, environments)

◆ Proposal that the Program could be presented to faculty with a judicious regard for academic freedom, coupled with a positive approach to "do the right thing"

Final Conditions for LSC Unanimous Support of LSSP

◆ Any/all science/research faculty would be invited to review/comment on a draft LSSP

◆ Participation in the Program would be voluntary

◆ Program would constitute a "guidance" approach, rather than a "formal policy"

Obtaining Unanimous Support of the Laboratory Safety Committee (LSC). The LSC Chair presented arguments in favor of an LSSP (also summarized in Table II).

Given several rounds of discussion, the LSC unanimously agreed to support a Laboratory Student Safety Program, provided the LSC adopt the approach summarized in Table II. One key item to note is that the LSC Chair pointedly used very few

"OSHA-related" references, regulations, requirements, or rationale to defend the need for (or, in fact, *develop*) this Program. On the contrary, EH&S wanted to stay clear of the natural resistance faculty have to OSHA (as a whole), and, moreover espousing regulatory or otherwise "bureaucratic" rhetoric to faculty OR students. And, while OSHA's laboratory Right-to-Know Standard served indirectly as a "template," the *single* focus, in this author's view, had to be student **health & safety.**

Stage II: Drafting the LSSP (September - November 1994)

Selecting and Utilizing Copyrighted Reference Material. The LSC began with a two-part format for the LSSP Manual: **Part One - LSSP Guidelines** (i.e., methods of incorporating safety instruction into lab course-work, and safety topics to include); and **Part Two - Laboratory Safety Guidelines** (i.e., elementary principles of laboratory safety, provided in order to assist inexperienced laboratory instructors). Immediately thereafter, the LSC opted to utilize excerpts from the American Chemical Society's 1990 Publication, Safety in Academic Chemistry Laboratories, with which to draft **Part Two** of the LSSP. In order to do so, OU (in Stage IV) provided the ACS Publications Department with a draft copy of the LSSP, and subsequently entered into an agreement with the ACS, wherein it stated in writing that a) the LSSP manual would be used internally only, and distributed with a corresponding letter prohibiting independent photocopying of the document; and b) that OU would notify the ACS whenever copies of the LSSP manual are generated, and pay (upon invoice) the ACS $1 for each printed.

Using the "Team" Approach to Drafting the LSSP. The LSC Chair chose to adopt a team strategy wherein ALL decisions were based on *unanimous* team support; (i.e., voting or other democratic strategies were not an option). Accordingly, four LSSP "drafts" were circulated to all LSC members for comments and/or formal "sign-off." Any major changes, issues or questions were brought up at LSC meetings, and a full consensus obtained before moving forward.

Stage III: Obtaining Departmental Input and Buy-In (Nov. 1994 - March 1995)

Presenting the Draft LSSP to Faculty and Staff. The Director of EH&S solicited "invitations" from Department Deans to present the Draft LSSP to their departments for comments and recommendations. Responses to these solicitations were varied, from no response, to declining, to accepting years later (see Table II). Subsequent responses to the LSSP Presentations themselves were equally as diverse (summarized in Table III).

Stage IV: Finalizing and Distributing the LSSP (April - September 1995)

Finalizing the LSSP. Based predominantly on the comments provided by the Department of Chemistry's faculty/staff, followed by full LSC consensus, the LSSP Manual was finalized. **Part One - LSSP Guidelines** provides the following: general expectations of Deans, Directors, Department Heads, Faculty, TAs, etc., with regard to implementing the LSSP; several methods of incorporating safety instruction into laboratory course-work; and recommended safety topics to include. **Part Two - Laboratory Safety Guidelines** provides (for benefit of the more inexperienced laboratory instructors) elementary principles of laboratory safety. The Table of Contents for Part One and Part Two are found in Tables IV and V (respectively).

Table III.   Departmental Responses to LSSP Presentations

Department	Invited EH&S to make an LSSP Presentation?	Departmental Response to LSSP Presentation
Biological Sciences	Yes, Immediately	Biology used time to discuss old (unrelated) funding issues. No LSSP input provided, but *did* provide opportunity to forge link between faculty and administrative agendas (which undoubtedly assisted indirectly with buy-in of the final document).
Chemistry	Yes, Immediately	Chemistry provided excellent recommendations, helping to make document more practical and workable at OU. Collaborative effort was critically instrumental to obtaining support from the Chemistry Department as a whole.
School of Engineering	Yes, but nearly two years later.	Some enthused; most *appeared* uninterested.
Health Sciences	No (SHS, in transition at time, had no response to EH&S)	NA
Physics	No (citing lack of laboratory student safety concerns)	NA

54

Table IV.   OU LSSP Contents - Part One: Program Guidelines

Section	Topic
I.	Introduction
II.	Scope of the LSSP
III.	Director/Department Head Responsibilities
IV.	Faculty Responsibilities:
A.	Implementing the LSSP
B.	OSHA requirements for laboratory "employees"
C.	Supervision
D.	Modeling
V.	Instruction Criteria:
A.	General
B.	Instructional formatting options. Examples: safety lectures projects SOPs video tapes
C.	Safety topics to include in safety curricula. Examples: laboratory protocol personal protection equipment laboratory supervision personal injury response fire prevention and response chemical handling/storage based on *hazard category* spill reporting/response/kits laboratory techniques for working with glass lab apparatuses and specialized laboratory equipment performing high hazard operations
VI.	Using EH&S as a Resource to Implement LSSP
VII.	Other Resources Available to Assist with Lab Safety Instruction

Table V.   OU LSSP Contents - Part Two: Laboratory Safety Guidelines

Section	Topic
I.	Laboratory Protocol:
A.	Housekeeping
B.	Personal hygiene
C.	Attire
D.	Cleaning glassware
E.	Transporting glassware
F.	Behavior
G.	Unattended operation of equipment
II.	Personal Protective Equipment
A.	General
B.	Eye
C.	Hand
III.	Admission to and Supervision of Laboratories
IV.	Personal Injury Response:
A.	General
B.	Chemical Spill to Body
C.	Chemical Spill to Eye
D.	Personal Injuries Involving Fire
E.	Other Types of Personal Injury
V.	Fire:
A.	Preparedness and Prevention
B.	In Case of Fire
C.	When the Fire Alarm Sounds
VI.	Laboratory Chemicals
A.	General Guidelines
B.	Chemical Hazard Categories
C.	Chemical Storage and Waste
VII.	Spill Reporting, Response and Kits
VIII.	Laboratory Techniques:
A.	Working with Glass
B.	Assembling Laboratory Apparatuses
C.	Using Laboratory Equipment
D.	Performing Hazardous Laboratory Operations
IX.	Laboratory Safety Equipment
A.	Eyewash Stations
B.	Safety Showers
C.	Fume Hoods
X.	"Sharps" Clean-up and Disposal
XI.	Material Safety Data Sheets
XII.	OU Chemical Hygiene Plan
Appendices	Condensed Laboratory Safety Rules
	Five Basic Laboratory Rules Poster
	Laboratory Safety Quiz
	OU Laboratory Safety Audit Checklist

Distribution of the LSSP. With the assistance of the LSC members, a profile of recipients was established, and corresponding names provided to the LSC Chair. Finally, following ACS copyright authorization (described in Stage II), the document was distributed to approximately 100 laboratory faculty, lecturers, TAs, graduate assistants, instructors and miscellaneous lecturers, in Biological Sciences, Chemistry, Engineering (Computer, Electrical/Systems, Mechanical), Health Sciences and Physics. Faculty responses ranged from indifference, to constructive criticism, to positive acknowledgment and praise.

Stage V: Quality Assessment (October 1995 - Present)

Development of a Quality Assessment Tool. The LSC Chair returned to the committee (following distribution of the LSSP), in order to propose a process which could objectively measure the effectiveness of the university's new LSSP. The Chair proposed that "Laboratory Safety Instruction Evaluation" forms be provided to all laboratory students (concurrent with the traditional course evaluation surveys). The purpose of this supplemental evaluation form would be two-fold: Provide the faculty and department heads an objective means of identifying areas where laboratory safety instruction may be lacking (*despite* the LSSP's attempts to the contrary), and provide TAs, etc. (via a review of the evaluation form at the *beginning* of the term) a "preview" of what would be expected of them, so that they could conscientiously work toward that end during the course of the semester. Therein, the tool for *assessing* the LSSP could double as a tool for *implementing* the LSSP.

Following review of a draft "Laboratory Safety Instruction Evaluation 'Boilerplate'", (designed by the LSC Chair for *chemical* laboratories), the LSC unanimously supported the utilization of the evaluation form -- under two conditions (both proposed by the LSC Chair): 1) Department Chairs (and their faculty if desired) would be instrumental in developing "customized" evaluation forms, which would be precise to the operations (and corresponding hazards) presented in each department; and 2) The faculty, Chairs, Deans and Department Heads would be at liberty to utilize the student feed-back as *they* saw fit, and would not be *obliged* to share the results of the evaluations with EH&S or the LSC.

In March 1996, the Director of EH&S met individually with Department Chairs to discuss their impressions of including Laboratory Safety Instruction Evaluations with their traditional course evaluation surveys, and to collaboratively create *customized* evaluation forms. The Chairs offered little to no resistance; in fact, the majority seemed encouraged by the prospect.

Implementation of Quality Assessment. Beginning in April 1996, Chemistry and Medical Laboratory Sciences began distributing their Laboratory Safety Instruction Evaluation forms. By December 1996, all science/research departments had distributed the evaluation forms to at least one semester of laboratory students. ALL Department Chairs agreed to share the results with the Office of EH&S, and in fact requested that EH&S analyze the results, and provide corresponding observations/recommendations.

Lessons Learned. A number of valuable observations have resulted from the Laboratory Safety Instruction Evaluations, both with regard to improving the evaluation surveys themselves, as well as laboratory safety instructional content/format. These are summarized in Table VI.

Table VI.   **Findings/Observations - Laboratory Safety Instruction/Evaluations**

Findings/Observations Related to Laboratory Safety *Instruction*

◆ Since medical laboratory courses offer what appear to be the most undisputed hazards, the evaluations in that department presented the most consistently positive responses

◆ Chemistry appeared (based on student comments) to rely too heavily on a single laboratory safety video, which is presented on the first day of class, and not substantially supported by "on-site" illustrations or demonstrations (e.g., locations and/or ***proper*** use of eyewash stations, etc.)

◆ Laboratory instructors ALL needed to be guided more specifically in the precise content of emergency response instructions, ***regardless*** of the level of hazards presented by each course, e.g., instructing students as to the following:
- laboratory room number (to provide to emergency response personnel);
- location of the nearest phone, inside **and** outside the laboratory;
- location of the nearest fire extinguisher, and guidelines for deciding *whether* to use it;
- location of the nearest fire exits; and
- the telephone number to reach the university's ***internal*** police department

---

Findings/Observations Related to Lab Safety Instruction *Evaluation Forms*

◆ Allowing the students an option of "NA" on one or more questions on the evaluation form presented a dilemma for EH&S to interpret (as it was unclear whether each safety topic *was*, in fact, "not applicable", OR whether the laboratory instructors failed to *illustrate* its applicability)

◆ Some terms and/or acronyms needed to be *explicitly* defined in the evaluation forms (e.g., "sharps" and "MSDS")

Current Status of OU's Laboratory Student Safety Program

In June of this year (1997), the Office of EH&S had its first evidence of how the LSS *Program* can be implemented. The laboratory coordinator/instructor for Oakland University's Undergraduate Fellowship Program in Biological Communication approached EH&S (after reviewing the OU LSSP), and asked the Office to provide a

58

Table VII.   Undergraduate Laboratory Safety Lecture Outline (based on OU
LSSP coupled with OU Laboratory Safety Audit Form).

GENERAL EMERGENCY PREPAREDNESS
Lab occupants know what room number/building they are in
Lab occupants know the locations of the: (campus) phone INSIDE the lab; nearest *campus* phone OUTSIDE the lab; nearest *pay* phone
Lab occupants know the telephone numbers for: ♦ OU Police - Emergencies   ♦ OU Police - Non-emergencies   ♦ OU Telephone Operators (Internal Directory Assistance)   ♦ Laboratory/Stockroom Manager   ♦ Environmental Health and Safety (EH&S)   ♦ EH&S Pager   ♦ Getting an Outside Line   ♦ Ameritech Directory Assistance   ♦ Reaching OU Police from Cellular Phone   ♦ Reaching OU Operators from Cellular Phone
Lab occupants know WHERE to find the nearest eyewash station and HOW to use it properly and **effectively**
Lab occupants know WHERE to find the nearest safety shower and HOW to use it properly and **effectively**
Lab occupants know WHERE to find the nearest Chemical Spill Kit and HOW to use it properly

FIRE PREVENTION/SAFETY
Lab occupants aware of **fire alarm pull station location(s)**
Lab occupants know location of **nearest fire extinguisher**; Date of last charge __
Fire cans (if present) contain fire blankets
All exits maintained to provide free and unobstructed exit
Lab occupants familiar with emergency response procedures (including **evacuation**)
Sign on door alerting fire fighters that lab contains > **2 gallons of flammables**
Emergency contacts and phone numbers posted on door and telephone
Storage of combustibles, e.g. cardboard boxes and paper towels is minimized

Table VII. *Continued*

CHEMICAL SAFETY - Flammable Chemicals
Flammable liquid containers do not exceed limitations for Class
⸕ 10 gal of flammable liquids are stored outside of **safety cans or cabinets**
⸕ 60 gallons of flammables in a **flammable storage cabinet**
Approved **"flammables" refrigerators** used for cold storage of flammable liquids
Flammables stored at least 18" **away from ignition sources**
Structural barriers separate **flammables from oxidizers and/or corrosives**
Flammables stored away from exits
Glass containers of **flammable** ..... chemicals (above 6 molar concentrations) **stored off of floor AND on shelves below five feet.**

CHEMICAL SAFETY - Safety by Hazard Category
Glass containers of ..... **corrosive** chemicals (above 6 molar concentrations) **stored off of floor AND on shelves below five feet.**
Chemical **containers labeled with hazards**; short-term (< 1 day) containers labeled with names
Highly/moderately chronic and highly acute **toxics labeled**; stored in **separate labeled areas**
Incompatibles stored separately (e.g., concentrated acetic acid or acetone near nitric acid)
Light reactives (e.g., hydrogen and chlorine) stored away from light
Water reactives (e.g., sodium, potassium, lithium) stored away from sinks/pipes
Peroxidizable chemicals (e.g., **ethers**) have dates of receipt AND opening; Containers were opened OR peroxide strip testing performed within six months of today's date
Picric acid is hydrated

CHEMICAL SAFETY - General
Chemicals stored to ensure that they are **resistant to sliding falls and/or spills**
Chemicals and food stored/handled separately; **no food waste** in lab trash cans
Haz waste is properly labeled and stored separately from other chemicals

Continued on next page.

Table VII. *Continued*

COMPRESSED GAS CYLINDERS
Gas cylinders properly **labeled** with the name of contents
Compressed gas **cylinders (whether full or empty) secured**
Flammable gas cylinders equipped with **flame arrestors**
Incompatible gas cylinders kept **segregated** (e.g., flammable gases such as acetylene or hydrogen near an oxygen cylinder)
Protective caps in place on cylinders which are not in use

SAFETY EQUIPMENT
Unobstructed **eyewash and safety showers**; labeled and identified
Eyewash and safety shower **flow tested** in past year (date of test _____)
Lab occupants know location of, understand how to use, and have **practiced using eyewash** equipment
Lab occupants know location of, understand how to use, and have **practiced** ("mimed" if necessary) **using safety shower** equipment
Biosafety cabinets properly functioning and **certified** in past year (date of certification
Goggles available at entrance to lab and provided to ALL visitors upon entering laboratory
Personal protective equipment available, used and in good condition (e.g., goggles, gloves)

GENERAL SAFETY/MAINTENANCE
Copy of *Chemical Hygiene Plan* available in laboratory
Laboratory **door locked when not in use**
Site-specific "training checklists" present, signed and up-to-date
Occupants are aware of location of (and procedure for obtaining) department's **first aid kit**
Spill kit(s) available in laboratory or employees know how to access during reg hrs and *after* hrs.
Laboratory occupants have been trained in proper **spill kit usage**
Hoses used for Bunsen burners are in good condition

Table VII. *Continued*

GENERAL SAFETY/MAINTENANCE (continued)
MSDS location(s) sign posted; MSDS or other safety literature available
Other signs properly posted (e.g., lasers, toxic chemical storage areas, etc.)
Sink drains clean and unobstructed
Generally good housekeeping (e.g., unobstructed aisles, floor clean and dry)
Appropriate waste receptacles for (non-biohazardous) **broken glass** provided
All **electrical appliances properly grounded**
Electrical **wires free from tangling**, obstructing aisles and corroding
Fans, pumps and motors have guards
Lighting levels provide **good illumination** in all walking, working and service areas

general laboratory safety lecture which would satisfy the content, and moreover the *intent*, of the LSSP. The lecture outline in Table VII (which used both the LSSP and the OU Laboratory Safety Audit Form as its "cursor") is the result. The subsequent 2-hour lecture (which expounds considerably on each of the items noted) was well received, both by the students, and the laboratory coordinator.

Moreover, OU's Department Chairs found their Laboratory Safety Instruction *Evaluations* (even those which were unfavorable) useful, and were openly appreciative of the "baseline" (and subsequent) information derived therein. Finally, a more interesting observation has been that it appears it is the Lab Safety *Evaluations*, rather than the LSS *Program* itself, which may be serving to motivate and modify laboratory safety instruction. This (hypothesized) scenario is depicted in Table VIII. Thus, while a *direct* "link" to the actual LSS *Program* cannot conclusively be determined, we have successfully promoted (however indirectly) the inclusion of critical laboratory safety instruction into laboratory curricula campus-wide.

Table VIII.   **The Use of OU's Quality Assessment Tool to Effect the LSSP**

Current and Projected Scenario

1) Faculty and TAs meet at the beginning of each term and review the Laboratory Student Safety Evaluation Forms (to be distributed to students at term-end);
2) In reviewing these forms, instructors make explicit preparations to incorporate this material into their curricula (in order to receive favorable evaluations at term-end);
4) Instructors **may** refer to *"Part Two - Laboratory Safety Guidelines"* of the LSSP Manual *(or* in fact some other reference literature altogether) to obtain information or guidance on topics with which they are less versed;
5) Instructors incorporate this material into their laboratory curricula;
6) Students obtain the necessary safety information;
7) Students rate the laboratory safety instruction favorably (on the Evaluation Forms) at the end of the term.

Future Goals of OU's Laboratory Student Safety Program

Based on the information received and analyzed thus far, the Office of EH&S shall offer the laboratory student safety lecture outline (found in Table VII), or its corresponding lecture, to future laboratory instructors upon request. Moreover, EH&S has indefinite plans to use the results of the Laboratory Student Safety Instruction *Evaluations* in order to: objectively analyze/interpret the student responses (and make corresponding recommendations to Department), improve the evaluation forms themselves (so as to ease and make more congruous the analysis thereof), and evaluate and modify on an on-going basis the Laboratory Student Safety Program.

Literature Cited

1. American Chemical Society Committee on Chemical Safety. **1990**. <u>Safety in Academic Chemistry Laboratories.</u> Fifth Edition. Washington, D.C.: American Chemical Society.

2. Oakland University Laboratory Safety Committee. **1995**. "Oakland University Laboratory Student Safety Program (LSSP)." Guidance manual developed for academic/research laboratory instructors, Oakland University, Rochester, Michigan.

3. U.S. Occupational Safety and Health Administration, Federal Register *52*, 31852 **(August 1987)**.

4. U.S. Occupational Safety and Health Administration, Federal Register *55*, 3327 **(January 1990)**.

5. U.S. Occupational Safety and Health Administration, Federal Register *56*, 64004 **(December 1991)**.

Model Chemical Hygiene Plans and Safety Programs in Academia

Chapter 6

Chemistry Safety at Michigan

Richard J. Giszezak

Department of Chemistry, University of Michigan, Ann Arbor, MI 48109–1055

Promoting health and safety in a medium-to-large academic chemistry department requires patience, persistence, and the efforts of many people. The University of Michigan Chemistry Department utilizes the efforts of the University Department of Occupational Safety and Environmental Health and a full time Safety Officer. Incoming graduate students are required to attend an orientation program that includes lab safety training and hands-on fire extinguisher training. The Safety Committee discusses and acts on appropriate departmental safety concerns. The Film and Video Library provides safety videos. Short chemical safety data sheets are provided in the teaching labs for all the hazardous chemicals used by the students. The Safety Officer and the graduate students assist with laboratory moves and clean-outs. Hazardous waste disposal is made as convenient as possible for all staff and students. Use of radioactive or biological materials brings on additional concerns and student training. Faculty incorporate safety concerns in lab experiments. Accommodations are made for students with disabilities. Promoting safety is a big job – everyone's job.

Promoting health and safety in a medium-to-large academic chemistry department requires patience, persistence, and the efforts of many people. The regulations of the Occupational Safety and Health Administration (OSHA) are mind-boggling alone. They include not only the Laboratory Safety Standard (LSS) regulations, but also the Hazard Communication Standard (HCS) or Right-to-Know (RTK) regulations, the Blood Borne Pathogens (BBP) regulations, and the Personnel Protective Equipment (PPE) regulations. The Environmental Protection Agency (EPA) promulgates the waste disposal regulations and the Department of Transportation (DOT) provides the shipping regulations. Other agencies whose regulations must be followed include the Nuclear Regulatory Commission (NRC), Departments of Public Health, and various other state and local agencies. And then there are various codes and standards that must not be forgotten, including fire protection codes, electrical codes, and ventilation standards. It is impossible for a health and safety professional to know all the regulations in detail, much less the faculty member, graduate student, or teaching assistant in a lab. Despite any objections some may raise, the point is to keep them all safe, even from themselves.

University of Michigan's Campus Safety Program

The University of Michigan's Chemistry Department on the Ann Arbor campus is an operation covering about 200,000 net square feet of space with 21 teaching laboratories and 120 research laboratories. It includes electronics and computer shops, a glass-blowing shop, an instrument shop with student facilities, a mainframe computing area, a crystallography facility, seven NMR's and a dozen other major departmental instruments. The teaching operation, with its own separate instrumentation, serves about 4500 students per year in laboratory courses. Personnel include 50 faculty and lecturers, 48 staff, 200 graduate students, and 30 research fellows. The loading dock facility receives over 800 chemical containers per month from over 40 suppliers. For the period July through December 1996, the Department shipped out over 16 tons (gross shipping weight) of hazardous wastes. And that does not include special waste roundup efforts.

Occupational Health at Michigan. The University of Michigan is fortunate to have its own Department of Occupational Safety and Environmental Health (OSEH). The Department serves the University community through several of its branches, which include a Radiation Safety group, a Biological and Laboratory Safety group, an Environmental Group, an Industrial Hygiene and Safety group, and a Hazardous Materials and Spill Cleanup group. OSEH performs fume hood monitoring and biosafety cabinet certifications for the entire University. They have produced many pamphlets and booklets summarizing OSHA and other regulations to assist departments in implementing the various programs. OSEH also acts as the official liaison between the University and all Federal and State regulatory agencies. Out of its budget OSEH covers such things as routine hazardous waste disposal costs, required OSHA physical examinations and respirators, basic prescription safety eyewear, many standard signs and labels, and baseline safety training sessions for employees throughout the University. OSEH also maintains a web site (1) with excellent safety and health resources available to all employees and students. OSEH continually adds to and improves the site.

Safety Management in the Chemistry Department. The Department of Chemistry has its own Safety Officer. The Safety Officer's responsibilities include chemical hygiene, safety systems and equipment, safety information such as Material Safety Data Sheets and signs and labels, general training, chemical inventories and disposal, safety inspections, safety emergencies, and physical exam and safety eyewear programs. Even though his title is the same as several OSEH employees, he is paid out of Chemistry Department funds. Close cooperation between the Chemistry Department and OSEH has greatly benefited the Chemistry

Department and also many other university departments. Chemistry was the first department to develop a comprehensive Chemical Hygiene Plan (CHP). The CHP grew into a very thorough safety book, including not only laboratory safety information, but also encompassing the Hazard Communication Standard, biological safety, radiological safety, machine shop safety, first aid information, hazardous waste disposal information, and much more to better serve the diverse departmental facilities. OSEH assisted with proofreading and provided many booklets as appendices. The CHP soon became the model for the whole University (and even a few local companies). OSEH has recently made a modified version available on its web site.

All new graduate students admitted to the Department of Chemistry and out of department graduate students who teach Chemistry courses are required to attend a three day orientation program that includes a session on the Laboratory Safety Standard and its application within the University of Michigan. They are also informed about other standard safety sessions that they may need to attend depending on their field of study and research. OSEH conducts training sessions on Bloodborne Pathogen safety and Radiological Safety regularly, and has a program to inform and assist individuals and spouses who work in hazardous areas and have concerns about reproductive health. The incoming graduate students also attend a session about fires and how to deal with them, including hands-on fire extinguisher training. They are given information on how to handle emergencies in the teaching labs, such as minor chemical spills, thermal burns, cuts, chemical exposures, or medical incidents (such as the occasional student who has not eaten in 24 hours and passes out at his or her bench). All of these efforts to create safety awareness, of course, are to encourage graduate students to promote chemical safety, as they teach undergraduate laboratories, recitations, and discussions. It is the Department's intention that at least some of the safety instruction provided will be passed on to the undergraduate students. The graduate student instructors who fail to adhere to or to make their students adhere to safety regulations may be written up with copies to the course instructor, administrative manager, and the associate chairperson for graduate studies.

Safety by Committee

One of the many standing committees within the Chemistry Department is the Safety Committee. The Committee is comprised of two faculty members, one lecturer, the administrative manager, the safety officer, and one graduate student. At their meetings they discuss any and all current departmental safety concerns and make recommendations where appropriate. They have been instrumental in upgrading safety eyewear to goggles in the teaching laboratories and in producing an in-house video about safety matters. All undergraduate students who do research for credit or pay are required to see the video which provides basic aspects of working safely in the research laboratories. They assist from time-to-time in inspections of Chemistry Department facilities, including both laboratory and shop areas. Close cooperation between Chemistry and OSEH has made these into joint inspections, checking for compliance with all the various regulations. The inspections are followed by detailed written reports and often by direct assistance in correcting problems. Reports are forwarded to lead supervisors and are used as official documentation of safety program compliance.

Safety Training

The University of Michigan Film and Video Library, in cooperation with the Department of Chemistry, has purchased a number of videos on laboratory safety topics, general safety topics, and regulatory matters. These videos are available to the entire University community. The Chemistry Department brings each pertinent video, a few at a time, to the Science Learning Center (SLC) for a one-week stay during semesters when classes are in session. The Science Learning Center is an inter-disciplinary study/resource center within the Chemistry Building, utilized by faculty, staff, and students.

All faculty are encouraged to and some faculty do include safety concerns in write-ups or oral discussions prior to undergraduate teaching lab experiments. In particular, one general chemistry course includes safety concerns in each experiment's instructions. These concerns are usually brought up in the discussion that immediately precedes the lab.

OSEH communicates through departmental safety representatives by newsletters, mail, e-mail, and occasional (four to six per year) short seminars featuring safety topics of current interest, including results of recent inspections. They frequently discuss what problems are occurring and what has worked for others. Their presentations aim to improve the safety atmosphere on campus among faculty, staff and students. They endeavor to turn many of the safety representatives whose safety duties are added responsibilities into true safety professionals.

Summaries of MSDS Information

Short chemical safety data sheets (2) are provided in the teaching labs for all the hazardous chemicals used by the students. The Safety Officer currently has computer access to a large database of Material Safety Data Sheets (MSDS)(3). Many other sources of MSDS's are accessible directly from our OSEH web site (1). Current plans include expanding access to the large database to a graduate student computing facility and an undergraduate computer facility (SLC), both within the Chemistry Building. One of our lower level courses frequently assigns a project to make students think about the manufacture, transportation, and use of chemicals in our society. Safety is always an important consideration in these projects.

Renovation Projects

The Chemistry Department Safety Officer also assists with laboratory moves and clean-outs. Moves can be brought on by retirements, research groups outgrowing their space, or the addition of new faculty members. Renovating older facilities can bring on major relocations and their associated headaches. During such moves, excess equipment, supplies, and chemicals are often donated to other groups who can use them. This is often accomplished through open houses or designated give-away areas, when recipients are not specified. Unwanted chemicals are often listed and posted for give away to help reduce chemical purchasing and disposal costs. The Graduate Student Council has occasionally sponsored a general clean-out day when any unwanted equipment, supplies of chemicals are taken to designated areas, from which anyone who can use them may take them. If a clean-out of a lab is called for, then everything but the furnishings must be removed. All drawers must be empty, postings removed, and bench tops and shelves wiped down. New occupants can not move in or renovations begin until formal decommissioning takes place, first by Radiation Safety personnel, if applicable, then by Laboratory Safety personnel. All this is to avoid saddling new occupants with safety hazards left behind by others, such as unknown hazardous chemicals.

Chemical Waste Management

Hazardous chemical waste disposal presents difficulties for any chemistry department. At The University of Michigan a Waste Lab at the loading dock takes much of the waste disposal burden off the researchers and teaching assistants. The Waste Lab receives empty waste containers (jars, bottles and buckets paid for by OSEH), affixes hazardous waste labels with our generator number, and places them in several closets and cabinets available to researchers throughout the building. Empty, labeled waste containers are taken directly to most teaching labs. Generators of the wastes must fill out the waste labels including chemical names and waste codes. Their initial lab safety training provides much of the knowledge needed to fill out the labels, their chemical hygiene plans provide more, and a single information sheet provides the appropriate waste codes. Preprinted waste container labels for each separate category are available to teaching assistants or students to affix to waste containers in the teaching labs. Each generator is required to set aside a designated hazardous waste accumulation area or areas. When generators are finished with waste containers (two months maximum accumulation period), they place them in designated locations from which Waste Lab personnel pick them up and take them to the Waste Lab. There the wastes are classified, manifested, and packed for weekly pickup by our hazardous waste hauler.

A color-coded plastic bucket system has evolved to help route various wastes to their proper disposal sites and minimize injuries to custodial staff and waste handlers. This system includes using red buckets for mercury-contaminated materials; yellow buckets for broken, non-contaminated glass; white buckets for other hazardous wastes; standard wastebaskets for ordinary trash; and blue buckets for spill clean-up materials. Students are informed of these distinctions at the very beginning of lab each term and throughout the term as needed since all of these buckets are used in each teaching lab.

Specific Handling Techniques

Use of radioactive materials is a special issue that is very highly controlled at The University of Michigan. Use of radioactive materials must be authorized by

Radiation Safety Service (RSS) who receive all radioactive materials, verify, survey, and log in the materials, and then deliver them to the proper labs. Though undergraduates seldom handle radiolabelled materials, they may work in a research lab where such materials are used, in which case the undergraduates receive appropriate training. Some sealed sources have been used in an upper level teaching lab. One of the lower level courses often uses a neutron source in a lecture demonstration of radioactive decay principles.

Work with biological materials is becoming much more prevalent in the Chemistry Department. Many of the materials present little hazard (Biosafety Level 1 materials). When it comes to the use of blood and blood products (Biosafety Level 2), Bloodborne Pathogen Standard training is required. A basic Exposure Control Plan can be downloaded from our OSEH's web site and modified to fit a group's particular needs. Hepatitis B vaccinations are offered to appropriate personnel. A booklet was developed to clearly spell out each area's responsibilities in the disposal of biologically hazardous waste. Undergraduate labs never use biological materials above a Biosafety Level 1 rating. They may encounter a higher level during work in a research lab and then they receive appropriate training.

Chemists with Disabilities

Accommodations are made as much as possible for employees and students with disabilities. All building facilities are wheelchair accessible. Teaching labs include select hoods and benches that can be used by those in wheelchairs. Recently, a special viewing devise was designed and built to assist an undergraduate researcher with severe vision problems in handling hazardous materials. The devise mounted on the front of a fume hood. It allowed the student to withdraw a chemical slightly from the hood for very close observation while still providing adequate airflow past the chemical and into the hood.

Summary

Do these efforts prevent all accidents, emergencies, and injuries? No, but hopefully they do eliminate some and help us cope with those that do occur, lessen their severity, and improve our recovery time. Promoting safety is a big job – everyone's job. Despite the costs of working safely, it pays!

References
1. University of Michigan Department of Occupational Safety and Environmental Health web site: http://www.umich.edu/~OSEH.
2. *The Sigma-Aldrich Library of Chemical Safety Data*, Edition II, Robert E. Lenga editor, 1988, used by permission.
3. *Sigma-Aldrich Material Safety Data Sheets*, CD-ROM English version with quarterly updates.

Chapter 7

Health and Safety Projects in the College of Chemistry at the University of California, Berkeley

Kurt W. Dreger and Steven F. Pedersen

College of Chemistry, University of California, Berkeley, CA 94720

The College of Chemistry Health and Safety Program (CCHASP) at the University of California, Berkeley, has been established to oversee the daily responsibilities of health and safety compliance. CCHASP is a progressive organization dedicated to ensuring that the College of Chemistry is a safe and healthy workplace. In response to the College's diverse research and teaching programs, CCHASP, has developed several programs specifically tailored for a chemical sciences and engineering research institute. Two of these programs, namely our laboratory safety seminars and comprehensive laboratory inspections, have documentable training features which compliment and thus reinforce each other. Health and safety training not only emphasizes an individual's responsibilities, but also those of a research group. Our undergraduate laboratory safety training program is also presented.

In today's workplace, the words health and safety are essentially inseparable. By creating the Occupational Safety and Health Administration, the federal government has seen to it that the nation's workers understand their right to demand a safe and healthy environment in which to work. The University of California at Berkeley takes OSHA's charge very seriously. The Office of Environment, Health and Safety on the Berkeley campus, is staffed with a variety of professionals dedicated to maintaining a safe and healthy environment for the entire campus population. One of this office's more "visible" clients is the College of Chemistry, a large teaching and research unit comprised of two academic departments, Chemistry and Chemical Engineering. Research and teaching facilities within the College include approximately 400 designated laboratory rooms and a variety of special support spaces, including several chemical stockrooms and hazardous waste collection facilities. Research personnel in the College number 800, consisting of principle investigators, graduate students and post-doctoral associates. In addition, nearly 2,500 undergraduate students are enrolled in laboratory instruction courses every semester. The College also employs approximately 200 professional staff in a variety of administrative, technical and facilities maintenance positions.

The diversity of research conducted in the College results in daily activities that include, but are not limited to, the use of radioactive materials, high powered lasers, biohazards, cryogens and toxic gases. Synthetic chemistry operations demand relatively large inventories of hazardous materials and reagents, all of which must be stored and used in a highly regulated manner. In addition, research and teaching activities produce hazardous and medical waste which must be managed by the College on a daily basis. Finally, large internal facilities and shops create a scope of health and safety issues comparable to small industrial settings.

In this chapter we review some of the health and safety programs that are currently being used in the College of Chemistry to address the above issues. It is our hope that sharing information in this format will foster collaborative efforts with other institutions that will allow us to amend existing and develop new programs that address chemical health and safety concerns in academic environments.

Health and Safety Organization

Establishing and maintaining a safe and healthy environment for all members of the College community is of tantamount importance to the University and College administration. Given the magnitude of potentially hazardous research activities and the level of EH&S regulatory concern, the College has invested its time and money in establishing a health and safety group, known as the College of Chemistry Health and Safety Program (CCHASP). This program is designed to compliment the university's Office of Environment, Health and Safety. Currently, CCHASP consists of 2 full time and 4 part time staff members who are dedicated to a variety of health and safety functions. CCHASP is involved in the following activities specific to the College of Chemistry:

- Daily hazardous waste management
- Safety training
- Laboratory & facility safety inspections
- Injury & illness prevention
- Industrial hygiene activities
- Facility renovations planning & oversight
- Emergency planning & response

The recent success of our program lies in a cooperative effort forged between CCHASP and the campus Office of EH&S. Even with a team of six full and part time staff, CCHASP relies extensively on the Office of EH&S, especially in the areas of information and technology sharing. The Office of EH&S is responsible for interpreting and conveying to the campus community all issues relating to regulatory compliance, handles all outside agency questions and inspections and oversees the chemical inventory program. Furthermore, the Office of EH&S handles all hazardous waste contract negotiations and operates the campus hazardous waste storage facility. This office also maintains a trained emergency response team that is on call to handle most incidents including those relating to chemical accidents. The Office of Radiation Safety serves the College of Chemistry by providing laser safety training and inspections and radioactive materials management.

College of Chemistry Health and Safety Projects

Laboratory Safety Training Program for College Researchers. CCHASP has developed a comprehensive safety training program for laboratory researchers. This program is designed to meet the following objectives:

- Prepare all researchers to perform laboratory work in a safe and environmentally responsible manner.
- Disseminate general health and safety information supplied by the Office of EH&S and discuss health and safety policies particular to the College of Chemistry.
- Meet all regulatory requirements for training of laboratory personnel (*1-2*)

The safety training program for laboratory researchers consists of two principle components. First, prior to being issued a key to his/her laboratory, all new researchers are provided an employee manual entitled "Who Does It!, Where to Find it!, and How to Do It Safely!". The "Who Does It!..." manual which is revised and reprinted once each year, contains a shortened version of the College's Chemical Hygiene Plan, Emergency Response Plan and a section on the safe use of a variety of common laboratory equipment. After signing that he/she has read and understood these three sections of the manual, the new researcher is issued keys, a College of Chemistry identification/purchasing card and his/her name is entered into a facilities database. With this database, the College is able to track each researcher's safety training records throughout his/her stay in the College. Noteworthy is a new safety feature of our identification cards, on which emergency response and MSDS information is printed. This card is used daily by all College of Chemistry employees and researchers because of its ability to serve as a key card for access to many facilities, photocopy machines, and purchasing terminals. This is one card everyone carries all the time.

The second component of laboratory safety training involves the mandatory attendance at a safety seminar titled "Laboratory Safety and Emergency Response for College Researchers". Attendance at the seminar is required for all incoming graduate student and post-doctoral researchers. The seminar is typically conducted five times each year; twice in August when all new graduate students arrive at the College and three additional times to accommodate new post-doctoral researchers who are hired throughout the calendar year at different times. The safety seminar runs approximately 3 hours. The first 45 minutes is devoted to a fire safety training session that is presented by members of the Office of EH&S fire safety group. This session includes a "hands on" demonstration of fire extinguisher use for those who volunteer. Although we do not require that our researchers actually practice using a fire extinguisher, we ask them to pick up a typical large size (~38 lb.) carbon dioxide extinguisher to provide them with a feel for the weight of such a device.

The remaining time consists of lecture based training on laboratory safety and emergency response procedures. Upon entering the seminar, each participant is handed a portfolio which contains a variety of handouts and pamphlets relating to all areas of laboratory safety. Much of this information is supplied by the Office of EH&S. Two handouts developed by CCHASP are: 1) "The Laboratory Safety Cheat Sheet"--a quick reference detailing how to comply with minimum health & safety standards; and 2) the "Researcher Information Checklist"--a listing of what the researcher needs to know prior to conducting research in the College of Chemistry. The second handout is also used in the CCHASP laboratory inspection program. The content of the safety seminar centers on explaining the researcher's personal responsibilities in maintaining a safe and compliant laboratory. In order to clearly communicate what a researcher's safety responsibilities are, we divide the seminar information into six logical categories which we feel help organize the information most efficiently. We refer to these six categories of safety information as "The Six Responsibilities of Laboratory Safety" and emphasize to researchers that by paying attention to these key areas, health and safety compliance can be attained in a relatively easy, yet responsible fashion. Table I defines "The Six Responsibilities of Laboratory Safety" and can be used to formulate an effective seminar outline for a university laboratory research setting.

Recognizing the need for continuous training in the area of health and safety, CCHASP is currently developing a new seminar series aimed at our research community. The objective of this component of training will be to provide advanced safety information on specific subjects relevant to researchers' needs. These seminars will be presented by guest speakers who are experts in a particular area of health and/or safety. Choice of speakers will be driven by input from the researchers. In addition to these research oriented talks, many other special topics seminars are offered by other departments including the Office of EH&S and Occupational Health Services.

Laboratory/ Facility Self-Inspection Program. The College of Chemistry is committed to maintaining a safe and healthy environment in all of its laboratories and facilities. The goal of the CCHASP team is to reduce laboratory and workplace related injuries and illnesses to an absolute minimum. We also look for novel methods of promoting an "attitude" in research and instruction that makes our

Table I. The Six Responsibilities of Laboratory Safety

Responsibility Category	Summary of Items Presented	Examples of Responsibilities
1. Safety Information Systems	Chem Hygiene Plan, Chemical Labeling Systems, MSDS	Know when & where to access information. Know how to interpret information.
2. Personal Protective Equipment (PPE)	Safety Glasses, Goggles, Face Shield, Gloves	Know what PPE to use and when it is required. Use PPE at all times when required. Maintain PPE.
3. Engineering Controls in the Lab	Fume hoods, Gas Cabinets, Solvent Still System Controls, Neg. Air System	Understand how these systems work. Learn to recognize when systems are not working. Understand system limitations. Report system malfunctions immediately.
4. Emergency Equipment & Response	Eyewash/ Shower, Fire Extinguishers, Pull Boxes, Evac Routes, Earthquake Preparedness, Chemical Spills, Personal Injury, Building Security	Test Equipment periodically and report malfunctions. Know details of emergency response plan. Minimize earthquake hazards in the lab. Know who to call when injured.
5. Chemical & Hazardous Waste Management	Container labeling, Chem Segregation, Transportation of Chemicals, Hazardous Waste Segregation, Hazardous Waste Container Labeling & Handling, Spill Trays, Gas Cylinders	Appropriate management of chemicals and hazardous waste is required at all times.
6. Housekeeping & Good Lab Practices	Maintaining emergency exits, Avoidance of clutter, Prompt Disposal of Materials, Electrical Safety, Buddy System, Rotovap Trapping Techniques, Injury & Illness Prevention Program	Eliminate excess clutter from lab. Maintain clear aisles at all times. Recognize electrical hazards and take steps to eliminate them. Never work alone.

community aware of their impact on the environment. A critical step in meeting these goals is the CCHASP Inspection Program, an internal, comprehensive self-inspection. The purpose of the program is to identify all college practices and situations that could result in personal injury, impaired health to our researchers, students, and staff and damage to facilities or the environment. The program also establishes a mechanism that ensures abatement of identified hazards in a timely manner.

Laboratory and facility inspections are carried out on a year-round basis. Advantages of this type of continuous system include: 1) providing the inspector more time to identify and *discuss* , with college personnel, any health and safety problems in a given laboratory or facility; and 2) staggering the workload of our College shops so that they may respond, in a timely fashion, to any problems identified. A complete inspection of a given laboratory group or facility within the college contains four phases:

Phase 1. Self-Inspection. The laboratory director or facility manager is contacted in writing and given the following inspection checklists:

- "Self-Inspection Safety Checklist for Laboratories (Facilities)"
- "Group Information & Training Checklist"

These checklists are designed to guide each group through a thorough self-inspection process that involves all areas of environment, health and safety compliance for their area. The self inspection checklist for laboratories currently consists of 66 items divided into several health and safety information categories including: chemical labeling, personal protective equipment, fume hoods, emergency exiting, earthquake preparedness, electrical, gas cylinders, lab equipment, and fire prevention and life safety equipment. Accompanying each checklist is an explanation sheet which articulates the health and safety requirements for each given checklist item. After conducting a thorough self-inspection, the checklists are forwarded to CCHASP for review.

Ideally, each lab or facility group completes the self-inspection process every year. *By allowing researchers to complete their own checklists prior to an actual inspection, we introduce a valuable element of training into the inspection program.*

Phase 2. Follow-up Inspection. After reviewing the self-inspection checklists for a specific group or facility, CCHASP schedules a follow-up inspection of the area. Problems identified in the self-inspection process (phase 1) establish the basis for an in-depth inspection of the laboratory or facility. The CCHASP inspector uses the same checklists used in the self-inspection phase to verify compliance with environment, health and safety requirements and to point out problems that may have been overlooked. An important part of this inspection program's success lies in the inspector's ability to offer immediate solutions to identified problems. The

CCHASP inspector brings a cart stocked with labels and spill trays to each laboratory or facility, allowing for immediate abatement of any labeling and segregation concerns. This "on the spot" abatement of recognizable hazards emphasizes CCHASP's commitment to solving problems, not just pointing them out. It has been our experience that this level of involvement often lays the groundwork for a good working relationship between researchers and CCHASP staff; one that is based on mutual respect for each other's professions.

Phase 3. Corrective Action Reports. In cases where noted deficiencies can not be immediately mitigated, CCHASP issues a report to the laboratory or facility manager which clearly identifies all remaining EH&S related problems. This report is issued within 2 weeks following the CCHASP follow-up inspection. The report explains any problems found during the inspection and suggests a course of action necessary to correct each problem. Such problems are subdivided into two categories based on the designated person(s) responsible for correcting said problems: 1) principal investigator and researcher; 2) College of Chemistry. College of Chemistry responsibilities include notifying any campus units that have domain over a particular area or problem.

Phase 4. Group Training Update. Following the completion of the inspection for a particular research group, the CCHASP inspector meets with the group to evaluate the inspection results and to refresh each group member's knowledge of environment, health and safety polices, procedures and practices. This type of closure meeting has tremendous benefits. First, it provides a mechanism for refreshing an individual's knowledge of safety requirements. Furthermore such a meeting serves to enhance a *group's* awareness of safety. Many laboratory safety issues are common to a group of researchers, not just an individual. Such an approach to training is too often ignored in academic settings, almost by default, due to the nature of academic research. That is, researchers are encouraged to think and act on their own. *Group cooperation in laboratory safety issues is the key to a safe working environment.* Indeed, individuals are responsible for their own actions at the lab bench. However, it requires a concerted effort to make sure that one does not get injured from, for example, a piece of common equipment. Another important advantage of these group safety discussions is that it allows the principle investigator to demonstrate his/her commitment to safety and to communicate his/her expectations of the group as a whole. Finally, the meeting offers another opportunity to document safety training in the College, *including documentation that the principle investigator has received such training.*

Undergraduate Laboratory Safety Training. The safety training and inspection programs described above are designed primarily for researchers, faculty and staff. However, as already mentioned, the College of Chemistry also opens its teaching laboratories to some 2,500 undergraduates per year. These students work in over 30 laboratories, all of which are subject to the same inspection process as described for research laboratories. More importantly, there is an obvious connection between our graduate student researchers and these undergraduates; the former are often the teachers of the latter. That is, the majority of laboratory instructors are graduate students. Therefore, by exposing our researchers to the different layers of safety training and inspections mentioned above, it is hoped that this "safety conditioning" will extend to their positions as teachers. To further solidify this connection between research lab and teaching lab safety, all laboratory instructors for our freshman and sophomore laboratories must attend an additional safety training session at the beginning of each semester they are to teach. This session starts with general safety issues, and then focuses on a safety checklist that the undergraduates are required to sign before checking into to their laboratory locker. The checklist consists of twenty seven items divided into five categories: Personal Safety, Fire/Earthquake Evacuation, Health/Injuries, Chemical Safety, and Equipment Safety. In addition to ensuring that the lab instructors understand all the items on this list, they are trained how to present this material to the students. For example, under the heading of Fire/Earthquake Evacuation, we require that the instructors take their students on a walking tour, pointing out the nearest fire alarm and emergency phone, as well as following the actual exit path the students are to take in case of an evacuation. Exiting information is of course posted in the laboratory and is also found in the students' lab manuals (3) in the form of building floor maps with arrows indicating the correct exiting paths. Other safety features found in these lab manuals include a discussion of MSDS's and how to access them electronically as well where to find hard copies. Furthermore, as a constant reminder of the importance of safety in the laboratory, clip art icons are placed in relevant sections throughout each experiment in the manuals.

Conclusion

In summary, CCHASP along with the Office of EH&S face daily challenges in a complex research and teaching environment. The programs described in this chapter are intended to be dynamic. We are interested in discussing these programs with others in the field and hope that the information provided above is of use to those involved in chemical health and safety.

Literature Cited

1. California Code of Regulations, Title 8, Section 5191: California OSHA Standard.
2. Code of Federal Regulations, Title 29, Section 1910.1450: Federal OSHA Laboratory Standard.
3. Chemistry 3A and 3B Laboratory Manuals. Steven F. Pedersen and Arlyn M. Myers. Department of Chemistry, University of California, Berkeley, CA.

Chapter 8

Electronic Media in Chemical Safety Education at University of California at San Diego

R. Vernon Clark and J. G. Palmer

Department of Chemistry and Biochemistry, University of California, San Diego, CA 92093–0303

A brief history of lab safety education since 1986 at the University of California, San Diego is presented along with the rationale for, and content of the various forms this education has taken. The use of electronic media in regulatory compliance includes environment, health and safety training, chemical inventory control, material safety data sheet repository and chemical hygiene plan documentation. A summary of present efforts and future goals is included.

The use of electronic media to convey science, and the attendant safety considerations dates at least back to March 10, 1876 when Alexander Graham Bell and Watson, located in different rooms, were about to test the new type of transmitter. Watson heard Bell's voice saying, "Mr. Watson, come here. I want you." Bell had upset a battery, spilling acid on his clothing.
(Lucid Interactive. **1997** http://www2.lucidcafe.com/lucidcafe/library/96mar/bell.html)

History of Laboratory Safety Education at UCSD

Prior to the wide spread use of computers, the safety information in chemistry courses was limited to what was expressed verbally or written in the text book, lab book or on supplemental information sheets describing the experiments. The safety information was ancillary to the material being taught and the emphasis given it varied widely depending on the course instructor. This is still true today. Often safety information is best learned along with the proper technique to perform an operation. A series of videos depicting laboratory techniques with safety information were created in the 1980's and made available to the students through the Playback Center at the Undergraduate Library. (**1997** http://orpheus.ucsd.edu/ugl/playback.htm) Many excellent references regarding hazardous materials management, hazardous waste management and laboratory safety are available still in the old-fashioned book form. (*1-6*)

Safety Seminars from 1986. Formal, safety specific seminars were instituted in 1986. Initially the current departmental safety director invited the campus safety training officer to develop a training session for the undergraduate lab students. For the first two years these lectures were given inside one of the lab rooms. Approximately eighty students would crowd into a chemistry lab designed for twenty-four people. An overhead projector would be directed toward a blank spot on the wall and the lecture would last for just under an hour. This would be repeated for each of the five lab class time slots for the three different beginning level courses each quarter. Three years later the mandatory safety seminars were shifted to a larger lecture hall outside of the regularly scheduled lab times. There were two opportunities for the students from each of the three lab classes to attend. This resulted in a total of six evening lectures. Attendance was taken by having students initial next to their name on a pre-printed roster. The material remained fairly general and broadly applicable but not truly specific to the operation of the labs. Once in the large lecture halls, the training began to incorporate elements of commercially available videos along with the overheads and live demonstration of the explosive power of a highly flammable, well-mixed fuel. After a few years, the instructor developed two custom training videos and used these along with the other materials. This schedule format continued through 1995.

Safety Seminars from 1995. In the fall quarter of 1995 the current environment, health and safety (EH&S) specialist hired specifically for the undergraduate teaching labs began co-teaching the evening sessions. At this time, more lab- and course-specific information was integrated into the presentations. With the two instructors, the presentation became somewhat less tedious for the students. Attendance was taken, as it is now, by having the students complete a short quiz that covered the material from both the presentation and a summary sheet which was given out. (VernonClark, R.; Palmer, J. **1997** http://www-chem.ucsd.edu/Courses/CoursePages/Uglabs/Education/organic.safety.exam.html and safety.lect.notes.html) With this form of taking attendance it was much easier to verify that each student was trained. The next change was to find room on campus to house about 250 students for five separate lab time slots with all three introductory courses. In essence we went back to holding the safety lecture during the lab period but in a lecture hall, not in the lab rooms. We also determined that the variation in information provided between the courses during the safety lectures was very minimal, so all of the course safety lectures could be held at the same time. Our current schedule allows us to train all of the students in five lectures over two days. The overheads for the lectures were generated using PowerPoint. (VernonClark, R.; Palmer, J. **1997** http://www-chem.ucsd.edu/Courses/CoursePages/Uglabs/Education/Presentations/ug97pp.ppz)

Abo··· a year ago we began to develop information for distribution on the world wide web.

The Purpose of Our Efforts

Occupational Safety and Health Administration (OSHA) Lab Standard that promulgates the Chemical Hygiene Plan [29 Code of Federal Regulations 1910.1450] and the Hazard Communication Standard [29 CFR 1910.1200] provide the guidance and direction we adopted in establishing our programs. There are several layers of persons associated with our educational facility and the interaction with each must be slightly different. The different types of people are the general public, undergraduate students, the undergraduate student employees, the graduate student employees, viz. teaching assistants, the non-student teaching assistants, the lab staff employees, the temporary faculty lecturers, the tenure-track faculty, the department and university administrative staff. The University administrative staff includes the campus Environmental, Health and Safety division employees.

Signs alert the public. Our priorities for minimizing the risk of harm to people is based upon the probability and severity of exposure. In the undergraduate teaching labs, we place signs on the lab and storage doors to alert people of the hazards. In the building where this activity is located all of the lab doors are exterior doors. For the public, this is the extent of our educational effort. In California, Proposition 65 requires a sign to be posted wherever carcinogens are used or stored. Besides posting these signs on the outside of each lab door, we also post a sign describing what type of lab is used in that room. Near the doors of the research labs is a NFPA diamond sign where the hazard information is intended to be placed. We rely upon the departments of the other university employees to train their own staff.

Developing OSHA Savvy Students. The instruction of students to be OSHA- and environmentally-aware has several motivations. Primarily the goal is to minimize the hazards and eliminate injuries to everyone in the lab. The desire to impart information that will develop a better citizen and a more marketable graduate are important as well.

While most of the students are not employees, they have a vested interest in knowing the hazards associated with the learning experiences in the lab. Currently we provide an hour long safety lecture and access to material safety data sheets (MSDS) and UCSD laboratory safety guides which include the chemical hygiene plans for each of the courses. We have rewritten the general chemistry lab manuals to both eliminate hazardous waste and more fully delineate the hazards present in each experimental step. In addition, for any self-motivated learner, many of the documents relevant to help students become OSHA savvy are available through the departmental web server. (VernonClark, R. http://www-chem.ucsd.edu/Courses/CoursePages/Uglabs/)

Employee Education. We annually educate all of the staff, including the departmental administrative staff, following the Illness and Injury Prevention Program (IIPP) and as prescribed in the UCSD Business plan training (**1997** http://www-ehs.ucsd.edu/iipp.htm and haztrain.htm). We have made available both CPR and basic first aid training to these folks as well. The undergraduate teaching lab staff, including the student employees, are educated in a slightly more focused fashion as they are intimately involved in carrying out procedures that will insure regulatory compliance. The ten or so student employees gain a much greater appreciation for the details of the compliance efforts. In addition to the lectures they sat through in the lab courses, they endure the IIPP training, initial individual employee health and safety orientation, training for specific tasks and on-going lab staff-wide training sessions. For the regular lab staff, and a few of the students, we have a CD-ROM based training, specifically developed for the University of California. ("UC LabSafe, Working Safely With Hazardous Materials in the Lab", by Quint, J. **1994**). When these student employees graduate they have the benefit of more finely honed health and safety skills.

The Challenge of Educating the Educators. One of the most critical groups of people we educate are the graduate students and faculty who are directly responsible for the activities in the both the research and teaching labs. In our system, the undergraduate teaching laboratory instructor may, or may not be a tenure-track faculty. Over the past several years, for the general, analytical and organic lab courses the faculty have been temporary lecturers with annual appointments for a maximum of three years. In the organic courses we have had four instructors in five years. Also for most quarters, there are about thirty-six laboratory teaching assistants with a variety of levels of experience and education. Most of the teaching assistants are department graduate students but due to the generally increasing numbers of students taking lab classes as illustrated in Figure 1, the department has had to draw from a wider pool of people. These include graduate students from other departments, recent bachelor degree graduates from UCSD and elsewhere, and even a few postdoctoral researchers. While at the beginning of the first year of graduate school the Chemistry and Biochemistry Department provides intensive safety training, many other teaching assistant employees have not had that experience. So, members of the lab staff and the undergraduate lab EH&S specialist meet with nearly all of the teaching assistants and the lab instructors at the beginning of each quarter, usually during the initial meeting each instructor holds with their teaching assistants.

The research lab specific safety training is primarily the responsibility of each principle investigator (PI). The safety director augments this education whenever the PI requests. This usually takes the form of a special research group meeting. During the course of a graduate student's education doing research it is highly likely that a variety of individuals will describe the safe operation of materials and specific equipment. In order to facilitate the documentation, for OSHA compliance, of this training the research labs are supposed to maintain a training log for each person.

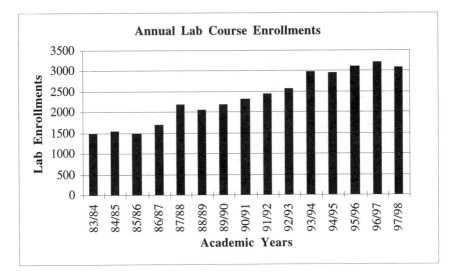

Figure 1. Increasing Numbers of Students Taking Lab Courses

Environment, Health & Safety Education Content

In educating people, a variety of approaches have proven useful. The information we generally pass along can be placed into a few categories. Theses include: preparing for using hazardous materials, using hazardous materials and equipment both environmentally responsibly and safely, preparing for emergencies and disasters, reacting to an emergency, the proper disposal of hazardous wastes and developing a regulatory compliance awareness. The content of the mandatory safety lectures given to the undergraduate lab students roughly correspond to these topics. The specific areas of personal protective equipment, hazard communication, hazardous materials, hazardous waste, sharps waste, fire, injury, earthquakes, equipment maintenance, and laboratory rules are presented to the students.

Perhaps the most effective and rapidly useful health and safety training we can provide is a safe method to carry out a specific procedure using hazardous materials. As previously mentioned, several different techniques have been captured on videotape and are available from the Playback Center inside the nearby Undergraduate Library. The topics covered include: column chromatography, extraction, fractional distillation, gas chromatography, infrared spectroscopy, melting point determination, nuclear magnetic resonance, recrystalization, simple distillation, thin layer chromatography, safety, solution preparation, spectrophotometric analysis, use of the burette, use of the pipette, and weighing. Also available are videotapes covering techniques for the microscale organic laboratory which includes an introduction, distillation, preparative gas chromatography, solvent extraction, infrared spectroscopy, simple crystallization, Craig crystallization, column chromatography and TLC, controlled atmosphere, and sublimation (these video recordings were produced by the Office of Instructional Development, University of California, Los Angeles in 1982). Another example of the electronic teaching of laboratory methods is the excellent work (7, 8) developed for biochemistry courses. This is a program, and a philosophy of teaching, that allows the students to explore as much of the background information as he or she needs to learn. This is now commercially available through ScienceMedia (**1997** http://www.ScienceMedia.com/).

The Use of Electronic Media to Approach Regulatory Compliance

The areas we use electronic media to help reach regulatory compliance include record keeping and document storage. The earliest use of the computers in our department for health and safety was in keeping track of the employees who had been trained in an annual safety refresher course. We also record the students who take the mandatory safety course each quarter in a database. We have more recently been tracking each of the employees who attend safety presentations on special issues, such as cardiopulmonary resuscitation and first aid. Many of the materials used during our undergraduate lab mandatory safety lecture are available on the undergraduate lab web site, including copies of the lecture notes and overheads.

Training. We are using the "UC LabSafe" computer-based safety training program on CD-ROM, that runs on the Macintosh platform, to supplement the in-house seminars for the undergraduate teaching lab staff. The areas covered in this training program are divided into the major classes of chemical, bloodborne pathogens and radioisotope hazards. Each of these areas has sections explaining how to identify hazards, protect against exposure, determine the best methods for waste disposal, obtain medical attention, and handle spills and emergencies. There are several computer based training systems and videos that include health and safety information available from commercial concerns. (These include Savant (800) 472-8268, Fisher Scientific (800) 766-7000 http://www.fisher.com, Lab Safety and Supply Inc. (800) 356-0783 http://www.labsafety.com and Coastal Video Communications Corp. (800)767-7703 http://www.SafetyOnline.net)

Chemical Inventory. In 1991, the campus division of Environmental Health & Safety made an Herculean effort to inventory and bar code all of the chemicals on the entire campus. This chemical inventory covered only those chemicals still in the manufacturer's original containers. For the undergraduate teaching labs this left out a significant amount of hazardous material that we maintain as a large number of custom solutions for specific experiments. Nearly fifteen months ago we began the process of re-training the lab staff, and selected faculty, to record both new purchases of chemicals and their consumption when the container is empty. This has proven to be the rate-limiting step in the implementation of a complete, current inventory recording system. The categories of information the campus retains in their chemical inventory database are: ownerlocation, pi#, piname, bldg, room, asset#, roomlable#, locationlable#, description (of space), asofyearmonthday, id, location, chemicalname, cas#, vendor, catalog, ordered (date), container (type), form (of matter), size, onhand, units (L, g, etc.), revised (date), and removed (date). Some of this information corresponds to information each of the lab staff wish to know as well; however, much of it is irrelevant to these users. We have developed a database using FileMaker Pro that allows a customization of the data presentation to more closely fulfill the needs for each of the lab staff. The kind information the staff generally wants to maintain relates to the course and experiments where the materials are used.

Material Safety Data Sheets. MSDS are kept both on paper copies and on-line through the undergraduate lab web pages. We recently created an index of all of the paper copies and are in the process of obtaining electronic versions whenever possible. One of our goals is to have each of the course experiments linked to the safety data sheet for the chemicals used in that lab. Our intention is to have our inventory list on-line and have each chemical name linked to the safety data sheet appropriate for it. Currently the best general MSDS on-line resource that we have found is through the Vermont SIRI web site (Stuart, R. **1997** http://siri.org/).

Chemical Hygiene Plans. CHPs provided the original impetus for getting on the web. Because the regents of the university passed the responsibility for OSHA and EPA compliance through the Chancellors, Deans, Department Chairs to the Lab Instructors, each course has its own chemical hygiene plan. In the undergraduate labs, the teaching assistants, and in some cases the instructors, change each quarter, so we revise each course CHP each quarter the class is taught. Because in our largest classes there are sixteen to twenty sections occurring at four, or five at a time in separate rooms, it is not possible for the instructor to be present in all of the lab sections. The site-specific information of our CHP contains the following sections: designated responsible individuals, room specific responsibilities, policies and procedures for receiving, storing and distributing chemicals, ordering and acquisition of chemicals, storage and maintenance of chemicals, peroxide detection tests, material safety data sheets and other reference materials available in the laboratory or stockroom, glove compatibility information, emergency response instructions, medical emergency response, laboratory specific procedures spill clean-up, site-specific engineering controls and ventilation information, personal protective equipment, writing standard operating procedures, chemical compatibility storage groups, and MSDS access on-line. A loose-leaf binder of the UCSD Laboratory Safety Guide, which includes the CHP, is kept in each lab room.

Computer Platforms

Several computer platforms are used in our efforts here. These include a mainframe Sun machine running Solaris that provides the department with a UNIX based operating system. The Sun is connected to a large number of dedicated x-terminals and a few text-only monitors. Additionally there are several Macintosh- and Intel- based personal computers networked in the department and on campus. Due in part to the wide variety of platforms used on campus we have developed a web site for the undergraduate labs to maintain current documentation and allow interested parties to keep informed. This system has allowed us to pursue the twin goals of approaching regulatory compliance and educating environment, health and safety-minded students, faculty and staff.

The initial use, for our undergraduate teaching labs, of the world wide web was to centrally locate many documents, including the lab course specific CHPs, and allow both easy access and updating when necessary. We became aware of the need to allow access to documents relevant to the students and staff, such as MSDS and course specific protocols. We are currently using the web for access to formatted documents, such as accident reports, spill clean-up procedures and hazardous waste labels. There is still a significant amount of work to be accomplished in this area. The safety section of the web pages is being developed to follow the form of a graphical tree to allow for easy topic recognition and rapid access to relevant information.

Summary of Present Efforts

We are building the web site to create a consistency in presentation and make it easier to access information. An on-line safety 'course' with an examination is being designed and may become a viable alternative for EH&S issues coverage in the more advanced courses. The chemical hygiene plans are being split into smaller parcels to allow for more rapid transfer over the network. The recently completed annual chemical inventory data is being entered into the database and this information will become available in report form for the lab staff over the web as well.

Future Goals

The ideal situation would have all of the standard operating procedures from all of the lab courses and lecture demos written and available over the network. We are interested in having all of the chemicals listed for each lab course, by experiment, on the web with hot links to their storage locations, material safety data sheets, relevant standard operating procedures and waste handling information. We are exploring the possibility of ultimately replacing the separate mandatory safety lecture with on-line information and examinations that could be integrated into the course. It is also necessary to make students, staff, and faculty aware of the external EH&S resources available both over the web and through other means. Our desire is to adequately educate students to be "OSHA savvy" and environmentally aware.

Conclusions

There are numerous challenges and opportunities to help students, faculty and staff to be more OSHA and environmentally aware via electronic media, particularly using the world wide web. For our academic institution, it is appropriate to pursue developing materials to produce the best students possible. This is especially true concerning environment, health, and safety issues.

Literature Cited.

(1)   Lewis, R. J. *Hazardous Chemicals Desk Reference, Third Edition*; Van Nostrand Reinhold: New York, NY, 1993.
(2)   NIOSH Pocket Guide to Chemical Hazards, U.S. Department of Health and Human Services, Public Health Service, Centers for Disease Control, National Institute for Occupational Safety and Health; U.S. Government Printing Office: Washington, D.C., 1994.
(3)   *Destruction of Hazardous Chemicals in the Laboratory, Second Edition*; Junn, G., Sansone, E. B.; John Wiley & sons, Inc.: New York, NY, 1994.
(4)   *Handbook on Hazardous Materials Management, Fifth Edition*; Cox, D.B.; Institute of Hazardous Materials Management: Rockville, MD 1995.
(5)   *Prudent Practices in the Laboratory, Handling and Disposal of Chemicals*; Committee on Prudent Practices for Handling, Storage, and Disposal of Chemicals in Laboratories, Board on Chemical Sciences and Technology, National Research Council; National Academic Press: Washington, D.C. 1995.
(6)   *CRC Handbook of Laboratory Safety. Fourth Edition;* Furr, A. K. Ed.; CRC Press: Boca Raton, LA, 1995.
(7)   G. Wienhausen and B. Sawrey, "Understanding Science: Visualizing the Molecular World and Simulating the Equipment," Proceedings of the International Symposium on Mathematics/Science Education and Technology **1994**.
(8)   G. Wienhausen and B. Sawrey, "Beyond Bio 101: A Report from the Howard Hughes Medical Institute," pp. 42-45 **1997**.

Chapter 9

Chemical Hygiene Plan Program at North Carolina State University

George H. Wahl, Jr.,[1] and David Rainer[2]

[1]Department of Chemistry, North Carolina State University,
Raleigh, NC 27695–8204
[2]Director, Environmental Health and Safety Center, North Carolina State
University, Raleigh, NC 27695–8007

ABSTRACT

In the mid 80s, few Universities were "OSHA Savvy;" only a few were even
"OSHA Aware!" As the national carcinogen scare spread, we began our
development of a program to insure safe handling of chemicals university-wide. As
we progressed, it soon became obvious that education was our major problem.
Researchers had been using questionable procedures for decades and had difficulty
understanding the need for change. Ultimately, we developed and reviewed over
eight hundred laboratory Safety Plans (equivalent to Chemical Hygiene Plans
(CHPs)). In the second phase of our Safety Plan program, we have introduced an
emphasis on process safety analysis as a safety triage technique for locating those
laboratories that will require extra attention and precautions.

BACKGROUND

North Carolina State University, with an enrollment of more than 27,000 students,
is the largest university in the state. As a land-grant institution, we have
traditionally been very strong in the fields of agriculture and engineering.
Additionally, we have developed world class programs in Textiles and Forestry, and
more recently have excelled in Physical and Mathematical Sciences as well as
Veterinary Medicine.

With all of these science and engineering programs, we have a very large number of
laboratories on our 600 acre main campus, on our new 1200 acre Centennial
Campus, and at various research and extension locations in each of our one hundred
counties. Thus, in 1984 when national concern over potential laboratory exposures
to carcinogens was very high, our Chancellor appointed a campus-wide committee

to develop a Carcinogen Policy. One of us (GHW) served as Chair of that committee for ten years.

After first determining that all university laboratory constituencies were represented, our second act was to change our goal. We determined that our focus was much too narrow. We therefore proposed to our Chancellor that we develop a Hazardous Materials Policy. This recommendation was quickly accepted and we set about our task.

As a committee, we quickly realized that evaluation and control of occupational and environmental hazards has often been thought to be within the span of control of the health and safety professional. However, our emerging campus paradigm, reached by committee consensus, affirmed that personnel working in academic laboratories, and principal investigators must recognize and accept responsibility; safety is not somebody else's responsibility. Yes, we generally use small quantities of chemicals. Yes, we are highly trained in our own areas. However, as a practical matter we use extremely hazardous materials and our processes change often.

Our committee also acknowledged that we as faculty do not employ the rigorous analysis and safety reviews necessary to help assure a safe laboratory. This is attributed to the changing nature of our lab activities, and the absence of a well developed "safety first" culture. Some of our students are novices at laboratory safety and they are not always closely supervised. Our facilities are not always the most robust because our buildings are aging, and money is often preferentially directed toward research, rather than toward engineered safeguards. Many facilities are "grandfathered" and cannot meet the latest code requirements that address laboratory safety and mandate the use of specific engineered safeguards for new construction.[1,2] Also, responses to some surveys seem to suggest that many faculty members lack the expertise, the background, the time or the incentive to integrate safety and health information into their courses.[3]

Our "Hazardous Materials Committee" met approximately every month for a year or more, discussing how best to overcome these conditions that seemed destined to keep safety improvements from becoming a reality. Between meetings, we made a point of discussing our task with colleagues. We quickly learned that we had a problem that was mainly an education problem. We had to "Train the Trainers." Most Principal Investigators had been educated by a previous generation of scientists for which Safety was treated as a "four letter word," if at all. "We've been doing it this way for twenty five years! Why do we need to change now?" was the most often heard comment on our committee's task.

FIRST SAFETY PLANS

The committee finally decided that each Principal Investigator (PI) (a person who is the PI of a grant or contract that supports much of the work being done in a given lab; or the person in charge of an instructional or service laboratory) needed to produce written procedures, specific for each laboratory, and agreed to by everyone working in that laboratory. The great diversity of activities on campus was inconsistent with a single University Safety Plan. Furthermore, by producing a custom plan, we hoped that each PI and the people working in that facility, would have a greater sense of ownership and be more likely to follow the plan.

We needed an unimpeachable outside reference source for safety plans to convince our colleagues that there was a better way to do business. We found this in "Prudent Practices" the two volumes developed by Blue Ribbon panels of scientists with unimpeachable reputations who took the time to develop better ways to deal with chemicals in the laboratory.[4, 5] We also studied and then adopted "SAFETY in Academic Chemistry Laboratories" (SACL), a manual we had been giving to Chemistry graduate students for some time, as an inexpensive source of sound safety advice, that we could distribute widely across campus.[6]

Further study of SACL suggested that it provided general guidelines for most of the typical chemical laboratory procedures that were common in our laboratories. Therefore, after generating a word processing document containing a "fill in the blank" form to create the "boiler plate" that surrounded the Safety Plan, we adapted prose from SACL, and from "Prudent Practices," and generated many typical safe laboratory procedures.

With this new document, the onerous task of producing a Safety Plan became much easier. The typical Principal Investigator (PI) need only open this document and add and subtract sections as appropriate for a given lab to produce a DRAFT Safety Plan which could then be perfected through conversations with those working in the lab.

As a couple of years passed, and the number of Safety Plans for different labs began to increase, one of us (GHW) was first given one month's Summer salary, and then a half time secretary to assist in this large task.

We began to receive some resistance from PIs. Several believed that if they wrote down the safe procedures that must be followed in their labs, their legal liability would be greatly magnified! One person was so sure of this position that he took an early retirement rather than prepare and sign a Safety Plan for his lab.

We also spoke with our Deans and convinced them that they really did have authority over, and therefore responsibility for, the labs of the PIs in their Colleges. Once this hurdle was crossed the number of plans increased at an accelerating rate.

At first, the (all volunteer, faculty and staff) Hazardous Materials Committee reviewed each Safety Plan as it was submitted. They usually provided not only general proof reading, but often significant directions and assistance with a significant revision. However, this very slow and onerous activity was replaced with a review by the Chair of the committee (occasionally assisted by one or more members of the committee) and then by the Director of Life Safety Services (now the Environmental Health and Safety Center).

The first round of Safety Plan development ultimately produced more than eight hundred different plans, and we believe we made a significant improvement in campus laboratory safety. It also heightened campus-wide safety consciousness and contributed to the building of a new Environmental Health and Safety (EH&S) building and the significant expansion of the EH&S staff.

REVISED SAFETY PLAN PROGRAM

Recent incidents [7, 8] and regulatory actions [9, 10] demonstrate that campus communities must be involved in a continuous review and improvement of their safety programs. In particular, a few incidents on our campus helped others to "see the light" and become more proactive with respect to laboratory and building safety.

The first incident happened in a semiconductor research and development laboratory. A fire occurred in a metal organic chemical vapor deposition (MOCVD) system in a laboratory of about 1200 sq ft. The MOCVD research apparatus used hydrogen, arsine, phosphine, and numerous pyrophoric metal organic compounds including trimethyl aluminum, trimethyl gallium, and dimethyl zinc. The exact cause of the fire was never affirmatively ascertained. However, the incident caused the total destruction of equipment in the lab as well as in adjoining labs from severe smoke damage. The total laboratory remediation, exclusive of lost time and lost data, and the fact that eighteen graduate students had no place to do experimental work for two years was approximately five million dollars!

The second incident happened in a laboratory hood constructed of polypropylene. These plastic hoods are frequently used in semiconductor research operations. An overheated hot plate caused a fire, limited to the hood, but caused damage that required an expenditure of $43,000 to repair.

A third incident involved the failure of a regulator on a compressed hydrogen gas cylinder. Piping attached to the regulator was not rated for full cylinder pressure and when the regulator failed, the pipe also failed. The ensuing fire destroyed a $25,000 gas chromatograph and was fortunately extinguished before spreading.

We believe that common issues led to all of these costly incidents. We have therefore developed a new and improved EH&S Safety Plan program.

We have retained the concept of individualized safety plans for each laboratory, prepared according to a common university format. We have further strengthened our reliance on "SAFETY in Academic Chemistry Laboratories." A copy is given to each PI as a plan is developed or revised, with the stipulation that every employee covered by the plan must read and become familiar with it. We emphasize that the actual plan will be the foundation for the training of all persons working in the lab.

Draft plans are reviewed by EH&S, Three signed copies of each final plan are produced. One copy is retained in EH&S, one is given to the Department Head of the area in which the covered lab is located, and the third copy is kept in the lab, available at any time to all lab workers. It is kept along with the lab's inventory, SACL, and the MSDS for that lab.

Our program is now clearly focused around accountability and responsibility. EH&S programs, structure, and requirements are now specified in faculty and staff handbooks, which are also available at the university web site. Every new employee receives at least two hours of safety training and is provided with a brochure that reinforces the following program elements:

Elements of the Environmental Health and Safety Program

NC State University's environmental health and safety program encompasses seven major elements:

1. The most important element is **you**. You are responsible for insuring that all your activities result in an acceptable level of risk to you, to the others present, to the general population, and to the environment.

2. Your immediate **supervisor** is responsible for keeping you apprised of the approved procedures and information related to the health and safety aspects of your activities.

3. Your **laboratory director, or in the case of work with radiation, the authorized user in charge**, holds the authority for work with hazardous substances, radioactive materials, or radiation producing devices. They are responsible to insure that all work is conducted with full regard for personal safety and health and in accordance with the laboratory safety plan, or the approved radiation use project.

4. Your **department head** is responsible for establishing and maintaining a work environment that fosters the appropriate health and safety procedures in all activities of the department.

94

5. Your **unit safety committee** is the group that audits the general physical condition and operations of the department and seeks resolution of recognizable safety and health problems and acts as liaison with the university Environmental Health and Safety Center (EH&S Center) in matters of health and safety.

6. The **Environmental Health and Safety Center** administers the health and safety programs of the university. The EH&S Center provides professional assistance and expertise to members of the university community in matters of occupational health and safety.

7. **Environmental Health and Safety Committees** provide input to the EH&S Center and its components Radiation Protection, Environmental Affairs, Health and Safety, and Industrial Hygiene. The voting members of these committees are drawn from the faculty, and staff and are recognized scientific experts in areas related to the respective committee's activity. These committees include the Occupational Safety and Health Council, Hazardous Materials, Radiation Protection, and Unit Safety Committees. The policies delineated in the handbooks were developed by consensus of a campus safety and health committee, facilitated by the Director of Environmental Health and Safety and represent one part of the administrative scope of the EH&S Center program. The policies also outline the University safety committee structure, EH&S Center authority, and the other elements necessary to provide administrative support to the EH&S programs.

Additional Campus Safety Programs

Closely linked to the defined administrative policies are key programs associated with the University hazard assessment effort. These programs include:

- Laboratory Safety Plans
- Laboratory Inspections
- Laboratory Exhaust Ventilation
- Exposure Assessment
- Facilities Modifications
- Process Hazard Review

One of the cornerstones of the EH&S program is the new, revised laboratory safety plan program. The goal of the revised program is to develop the best possible understanding of the operation of each lab. The results of lab inspections and hood surveys, and reviews of facility modification requests from laboratories all go into an EH&S data base designed to facilitate evaluation of sites of high potential hazard.

Likewise, the evaluation of each safety plan provides further data essential to establishing a hazard rating for a laboratory. As a part of the safety plan review process, each PI must complete the laboratory process description form, modeled after 29CFR1910.1450(e)[11] (see Figures 1,2), and submit an up to date copy of the laboratory chemical inventory to facilitate the ranking of labs into a hazard class. This ranking is a subjective assessment of the degree of hazard to personnel and the environment presented by the procedures and chemicals in the lab.

A walk-through survey of labs which receive a high hazard rating is conducted. This survey provides an accurate evaluation of the need for personal protective equipment (PPE), and other safety devices. Follow-up is conducted as appropriate. During the walk-through, the lab tenants must be present since one of the survey goals is to interact with the investigators and to evaluate potential hazardous exposures. This process is summarized below.

Proactive Approach to Exposure Assessment

Review laboratory Safety Plans
- Chemical Inventories
- Process Descriptions

Rank laboratories according to hazard
Conduct walk-through of priority labs
- interact with investigators
- evaluate need for PPE
Follow-up sampling as appropriate
Establish personal contact
- provide information and feed-back
as needed, such as PPE selection
Document survey results
Forward recommendations to PI

Another key element tied to the safety plan review is the process hazard review (PHR) program.

PROCESS HAZARD REVIEW (PHR)

Regulations requiring the use of process hazard review techniques are being promulgated by various organizations. Among others, the U. S. Department of Labor's Occupational Safety and Health Administration (OSHA) promulgated a "Chemical Process Safety Standard" (29CFR1910.119) which became effective in February 1992.[12] This standard requires an organization to establish a process hazards management program if they use "highly hazardous chemicals" exceeding specified threshold quantities. To accomplish this goal, "process hazard review"

96

Figure 1

PRINCIPAL INVESTIGATOR: _____
BUILDING: _____
ROOM NUMBER: _____

1. Process	Metal Organic Chemical Vapor Deposition (MOCVD) system.
2. Hazardous Chemical/Chemical Class	2% arsine in hydrogen, 15% phosphine in hydrogen, 1.5% silane in nitrogen, 100% ammonia, 2 hydrogen cylinders, various organic solvents, small quantities of pyrophoric liquids including trimethylaluminum, diethyl gallium and triethylzinc, various photoresists.
3. Potential Hazards	Arsine and phosphine are acutely toxic gases with TLV's of 50 ppb and 150 ppb respectively. These gases do not have adequate physical warning properties. All hydrogen-containing mixtures are extremely flammable. Photoresists contain toxic and mutagenic solvents. Pyrophoric liquids ignite on contact with air.
4. Personal Protective Equipment	Self-contained breathing apparatus is required during specific maintenance procedures and during cylinder changes. Consult standard operating procedures available in the laboratory. Lab coats and safety glasses shall be worn in the lab at all times. Maintenance operations may require additional personal protective equipment. Consult laboratory standard operating procedures.
5. Engineering and Ventilation Controls	All gases are maintained in ventilated storage cabinets. The laboratory contains area monitors for toxic and flammable gases. Immediately evacuate upon alarm. Scrubbers are used to remove reactor effluent containing residual toxic gases.
6. Special Handling and Storage Requirements	This laboratory is approved for a total aggregate of 400 cubic feet of toxic and flammable gas. Gases are required to be maintained in ventilated cabinets.
7. Spill and Accident Procedures	Evacuate immediate and adjoining area upon alarm. Notify the Environmental Health and Safety Center. Small quantities of spilled organic solvents may be cleaned with activated charcoal or other appropriate spill absorbent material.
8. Special Animal Use Precautions	NOT APPLICABLE
9. Decontamination Procedures	NOT APPLICABLE

Figure 2

SAMPLE: STANDARD OPERATING PROCEDURE FOR HAZARDOUS CHEMICALS

1. Process	DNA extraction/recovery/precipitation
2. Hazardous Chemical/ Chemical Class	flammable liquids-alcohol, ether, acetone poisons/carcinogens-chloroform corrosive-phenol
3. Potential Hazards	fire; inhalation of toxic vapors, absorption through skin, eyes, and transport of toxic substance dissolved in them into the body; poisons can increase the hazard in any fire due to inhalation of volatilized reagent; chloroform vapor is toxic and damaging to tissues, absorption through skin; phenol is corrosive and direct contact can cause serious burns, vapor is also corrosive to tissue.
4. Personal Protective Equipment	safety glasses; gloves resistant to these materials (nitrile, for example; full-buttoned front or back closing lab coat.
5. Engineering and Ventilation Controls	all handling, pipetting, dilutions of these reagents must be done in a chemical fume hood; tubes containing these materials can only be removed from fume hood for vortexing or centrifuging if tightly capped; always point opened test tube or eppendorf tube away from face so as to minimize inhalation; be sure that tube cap is on securely before mixing.
6. Special Handling and Storage Requirements	avoid ignition sources such as hot plates, heat lamps or Bunsen burners; use metal tray or absorbent materials to contain any spilled solvent; never heat directly or work near a spark source; dispose of waste or used ethanol, phenol and chloroform in hazardous waste containers in lab hood; be sure that centrifuge and refrigerator can be used safety - that they don't generate enough of a spark to ignite reagent.
7. Spill and Accident Procedures	Needed: absorbent materials (vermiculite, paper towels if you have nothing else on hand) brush and dust pan, plastic bags, sealable can to contain clean up materials, protective clothing, gloves, safety glasses, and mask to cover nose and mouth if needed. For minor spills: open windows or ventilate area if spill occurs outside hood; cover liquid with absorbent material; brush into dust pan slowly and place in plastic bag; do not breathe dust from absorbent, place all clean up materials in plastic bag, then into sealable container; clean the area with water several times, wipe up, then soap and water, discarding materials as above; contact EH&S for removal of seal container, do not place clean up materials in regular waste.
8. Special Animal Use Precautions	Not applicable
9. Decontamination Procedures	Not applicable

(PHR) is required. However, even small quantity chemical users not within the regulated community will benefit and enhance the safety of work operations by employing PHR techniques. The goal of the NC State University program is to help assure that single point process or system failures do not cascade and result in catastrophic system or process failures.

PHR Program Goals

Assure that single point failures do not
    result in catastrophic failures
Increase use of self-analysis
Documentation
EH&S Center coordinates and participates in reviews
Students are trained in PHR techniques and
    become accustomed to generally accepted
    practices for the use of hazardous materials

The elements listed in the OSHA standard provide a suitable framework to begin the hazard assessment of a project and should be viewed as the minimum criteria utilized to evaluate process hazards. Many of these elements would appear to be good common sense; however, they are often overlooked because the assumption is made, incorrectly, of course, that nothing will go wrong. The hazard analysis is necessary to assess the broad range of hazards inherent in any high technology process, develop a consensus of a broad spectrum of experts, develop hazardous materials handling procedures and to establish guidelines for handling an emergency condition and process effluent.

There are many hazard analysis formulations which may be used effectively to assess process hazards. These include fault-tree analysis, failure mode and effect analysis (FMEA), what if analysis, hazard and operability analysis (HAZOP), check list analysis, and safety review among others. The specifics associated with these analyses can be reviewed by consulting the appropriate American Institute of Chemical Engineers Center for Chemical Process Safety references.[13]

The important point in effectively implementing any of these procedures is to assemble a multidisciplinary team (review committee) represented by a technical representative of the PI, along with safety, operations and/or other specialists or personnel as appropriate. Each representative brings knowledge and expertise of the process technology under review and provides differing perspectives of the hazards or problems to be assessed.

Many different factors need to be evaluated when determining which process hazard review methodology should be utilized. These include the complexity of the equipment or process, previous experience of accidents or incidents ("near misses"), and the potential consequences of an adverse event such as serious injury or community contamination. The complexity of analyses varies widely due to time available, cost, etc. In general, "what if" is the simplest and least expensive type of analysis, and the analysis used at NC State.

The important point in effectively implementing any of these procedures is to assemble a multidisciplinary team (review committee) including a technical representative of the PI, along with safety, operations and/or other specialists or personnel as appropriate. This is sometimes a difficult process to implement as it represents a change in culture. It requires the PI to seek affirmation of the safety of her/his process.

After assembling the PHR committee, the goal should be to review the hazards of a process in its entirety and to provide written documentation of the review. The documentation should include what elements of the system were evaluated, how the assessment was conducted, and most importantly, a description of any corrective action that needs to be taken to correct deficiencies. The committee should function as a team and include expertise in engineering, safety, and other disciplines as required.

Minimum elements of the OSHA standard which should be included in any PHR analysis:

- Identify previous incidents

- List applicable engineering and
           administrative controls

                   assess consequences of control failure

                   evaluate range of possible employee
                   safety and health effects

- Develop written operating procedures
           including plans for management of wastes

CONCLUSION

The goal of our program is to make sure that a comprehensive assessment of laboratory hazards is conducted to help assure the safety and well being of our faculty, staff, and students. While students are not specifically covered by the Lab Standard (they are typically not "employees"), lab safety plans are required for all instructional labs that employ chemicals, and the plan must constitute the basis of the safety training supplied to the students in that lab.

Implementation of this program has required a cultural change. Faculty are encouraged to seek outside review of their processes, and to accept responsibility for their operations. However, implementation of these programs has improved the safety of our operations and has heightened awareness of safety concerns and issues among the faculty and staff.

REFERENCES

1. Southern Building Code Congress International, Inc., Birmingham, AL.

2. Acorn, William, Code Compliance for Advanced Technology Facilities: A Comprehensive Guide for Semiconductor and Other Hazardous Occupancies, Noyes Publications, Park Ridge, NJ, 1993, ISBN 0-8155-1338-0.

3. Phillips, T.R., Occupational, Product, and Public Safety and Health Education: Components in ABET - Accredited Engineering Programs. Accreditation Board for Engineering and Technology, Inc., 111 Market Place, Suite 1050, Baltimore, MD 21202. 1994.

4. Prudent Practices for Handling Hazardous Chemicals in Laboratories, National Academy Press, Washington, D.C., 1981 (now available as Prudent Practices in the Laboratory, 1995, ISBN 0-309-05229-7.

5. Prudent Practices for Disposal of Chemicals from Laboratories, National Academy Press, Washington, D.C., 1983.

6. SAFETY in Academic Chemistry Laboratories, American Chemical Society, Washington, D.C. (various editions) ISBN 0-8412-3259-8.

7. Laboratory Fire Exacts Costly Toll, Chemical and Engineering News, June 23, 1997.

8. Barry, Michael, M.D., Brief Report: Treatment of a Laboratory-Acquired Sabia Virus Infection, New England Journal of Medicine, August 1995.

9. Toxic Dispute Costs Stanford $1 Million, Science, Vol. 266, October 14, 1994.

10. Boston U. Settles Environmental Charges, NACUBO Business Officer, November 1997.

11. 29CFR1910.1450 Occupational Exposure to Hazardous Chemicals in Laboratories.

12. 29CFR1910.119 Process Safety Management of Highly Hazardous Chemicals.

13. Guidelines for Engineering Design for Process Safety, Center for Chemical Process Safety of the American Institute of Chemical Engineers, New York, NY 1993.

Chapter 10

Safety in the Chemistry Curriculum at Iowa State University

Gordon J. Miller[1] and Paul Richmond[2]

[1]Departments of Chemistry and [2]Environmental Health and Safety, Iowa State University, Ames, IA 50011

At Iowa State University we have developed and are continuing to design programs into our undergraduate and graduate curricula that will establish a sound practice of proper chemical hygiene in the laboratory. Our efforts include: (1) for all entering graduate students, an introductory graduate course to teach proper laboratory procedures and handling of chemicals; (2) for advanced graduate students, a chemistry safety committee which interacts with our university's EH&S department to learn about EPA and OSHA requirements so as to maintain a high level of safety awareness throughout our research labs; and (3) for our undergraduate introductory chemistry students, supplementary readings and pre-laboratory quizzes designed with chemical safety in mind, e.g., waste handling procedures, first aid, MSDS's, and the use of safety equipment. The fundamental goals of our approach are not only to teach students appropriate safety measures, but also to involve them in the enforcement of basic prudent practices in the laboratory.

According to *Prudent Practices in the Laboratory*, "the new culture of laboratory safety implements the priority of 'safety first' through a greatly increased emphasis on experiment planning." (1) Of course, this attitude must be instilled early and developed continually, for it affects all members of the laboratory community. In many universities, safety education in chemistry has been delegated primarily to a few regulatory documents at the beginning of a laboratory course, or an occasional warning in the description of a specific experiment. Moreover, safety issues are seldom raised in lecture-based chemistry courses. As one of the first land-grant colleges in the United States, Iowa State University adopts an educational philosophy that emphasizes applied and basic research and blends practical with liberal education. Therefore, Iowa State has an orientation toward science and

101

technology, with strengths in engineering and agriculture. Chemistry, the central science, plays a key role in the curriculum of many students, many of whom will go on to scientific leadership positions in industry, academia, and government. Not only will these individuals carry scientific research into the 21st century, but they must also confront the responsibility of experimental implications and risks as well as the increasing regulatory actions designed to minimize hazardous situations.

At Iowa State University we have begun to implement programs into the undergraduate and graduate chemistry curriculum that will increase the awareness and knowledge of chemical hygiene and laboratory safety issues among our students. These developments impact both teaching and research laboratories. Furthermore, with the assistance of the Ames Laboratory, postdoctoral research associates also undergo continued education in proper chemical hygiene and safety. This chapter summarizes our attempts to educate our students about prudent practices in the laboratory, management of chemicals and chemical waste, regulatory agencies and their roles, as well as appropriate actions and responses when a laboratory accident occurs.

Safety in the Curriculum: When and How?

Chemistry faculty, staff, and students, as well as officials in the department of Environmental Health and Safety (EH&S) universally agree that safety education is of paramount importance in the scientific curriculum. Such ideas need to be fostered at the beginning and continuously redressed and repeated throughout the entire education. Since Iowa State University offers undergraduate and graduate degrees, there are two types of efforts required. At the undergraduate level, education of fundamentals is needed, which includes laboratory protocol, use of safety equipment, familiarity with material safety data sheets (MSDSs), basics of first aid, and some of the specific terminology. For graduate students, this education must be reinforced and reviewed. Moreover, since these students will go on to accept scientific leadership positions in industry, academia, and the government, awareness of regulatory enforcement and knowledge of management's responsibilities are additional skills that laboratory supervisors, faculty, and administrators require for the effective and proper management of their scientific laboratories.

In order to effect a sound education in proper chemical hygiene and prudent practice, total commitment by the faculty is demanded. This commitment starts in the research laboratory, in which graduate students and post-doctoral associates are performing research for several years. Faculty members must take a proactive (and not reactive) stance toward chemical hygiene, which includes written standard operating procedures, training records, safety reviews, inventories, and occasional laboratory inspections. These practices can also become part of the supervision of teaching laboratories. Since graduate students serve as teaching assistants for large laboratory courses, undergraduate students are learning most laboratory techniques and chemical hygiene from these instructors. Such habits include maintaining a well organized and complete notebook, sound planning strategies and pre-lab preparations, wearing safety equipment (e.g., goggles and lab coat), and appropriate disposal practices. Therefore, a training program for teaching and research assistants must be in place. Our efforts in the chemistry department have greatly benefited by close collaboration with both Iowa State University's Department of Environmental Health and Safety, which is headed by Emery Sobottka and contains a staff of approximately 30 members, as well as the Ames Laboratory (supported by the Department of Energy and operated by Iowa State University), which has its own EH&S division. The ISU EH&S department handles issues involving general safety, hazardous waste, radiation safety, environmental safety, industrial hygiene, and risk management throughout the university. This department works with both our research and teaching efforts by helping faculty, staff, and students create and maintain a chemically and physically safe working environment, as well as helping us to better understand and comply with OSHA and EPA regulations.

Teaching and Enforcing Safety Practices

Our efforts to implement safety education in the chemistry curriculum are directed toward both our undergraduate and graduate programs. The impact and the effectiveness of our programs still await review and assessment. Nevertheless, we believe that this continuing proactive direction will help maintain proper and prudent practices in our research laboratories as well as promote sound safety education in our teaching laboratories. Since many of our graduates move on to industrial or academic positions, we expect and sincerely hope that these ideas will create positive attitudes towards chemical hygiene and safety in research and teaching laboratories throughout the country. Table I summarizes where our students have gone upon graduation for the last three years. Records in the chemistry department allow a yearly assessment, whereas the numbers for chemical engineering majors are summed over the three-year period. Certainly, any safety instruction will lay the foundation of sound chemical hygiene at a wide range of working environments for our students.

The Undergraduate Program. Many undergraduates participate in one year of general chemistry, which includes lectures and weekly laboratory exercises. Students interested in pursuing life sciences, chemical engineering, and chemistry also pursue organic and physical chemistry. Furthermore, chemistry majors are required to perform an independent research project with a faculty member in the chemistry department. During the past academic year (1996-97), we have focused our efforts on the general chemistry program, because any safety instruction that takes place with these courses will have the broadest impact, and will allow us to evaluate the perceptions of a wide range of student interests. In order to gauge the potential impact of these programs, the final enrollment numbers for our general chemistry courses during the past three academic years are: 1994-95, 4254; 1995-96, 4211; and 1996-97, 3949.

Table I. Placement of Chemistry and Chemical Engineering Graduates from Iowa State University during Last Three Years

Students	Year	Number	Industry	Academic	Other
Undergraduates	1995-96	18	9	5	4
(B.S., B.A.)	1994-95	23	9	11	3
	1993-94	18	12	5	1
Graduates	1995-96	37	10	17	10
(M.S.; Ph.D.)	1994-95	42	17	15	10
	1993-94	47	17	22	8
Chem. Eng.	1993-96	423	183	90	150

Introductory Chemistry. Safety instruction was directed at the laboratory courses. During the past year, students were assigned sections from the book, *Working Safely with Chemicals*, C. Gorman, Editor and published by Genium Publishers, with each laboratory experiment. (2) Sections from this book were selected based upon an instruction of fundamental terminology and concepts, or a selection of passages which was relevant to the experiment at hand. Topics from this book that were emphasized included routes of entry, laboratory protocol, exposure limits of various metal and nonmetal ions, basics of first aid, and the MSDS's and chemical profiles. Students were expected to answer a short pre-lab quiz which examined whether they had read this material or not; these scores did contribute to the final semester grade. We found this text offered much relevant information, but the writing style created a barrier towards how seriously the students treated this material. Some of the prose involves a cartoon character's conversation with the reader that carries on throughout the book. Although this style may be engaging for some readers, it did not attract or motivate our students. Another exercise that is reviewed periodically throughout each semester is to have each student draw a plan of the teaching laboratory that points out all safety features. These include exits, routes to exits, telephone, fire protection equipment (blankets and extinguishers), showers, spill kits, first aid kits, etc.

Advanced Chemistry Laboratories. At this point, safety education is explicitly limited to brief safety-related descriptions included in each set of experimental instructions. In organic chemistry laboratories, students encounter hoods for the first time, and the extent of instruction depends primarily on the initiative of the teaching assistant in charge of a given laboratory.

Lecture Courses. At Iowa State University, the lecture is the place where each instructor can establish the greatest control over the education and perception of proper chemical hygiene. Speaking from our own personal experiences, safety education enters the lecture through demonstrations, anecdotes, and carefully chosen homework problems that combine fundamental chemical concepts with safety issues. If, as an instructor, you are able and willing to perform frequent chemical demonstrations, set the example by wearing goggles and a lab coat. While discussing the chemical principles in view, mention the planning and preparation that is needed to build the demonstration and comment on fail-safe measures that are available in case something should go awry. For many students, the instructor is the first professional chemist they will meet. Use personal anecdotes not only to relate chemical concepts, but also to share your own experiences involving chemical hygiene in the laboratory. Certainly the most memorable stories often involve an accident; what measures were taken by emergency personnel and how was the chemical problem dealt with? Finally, use or create homework problems for your students that address safety or even environmental issues: dimensional analysis problems are particularly appropriate here because fundamental units of measurements can change from one field to another. Thus, exposure levels for hazardous gases, like H_2S or CO, are often quoted in ppm, but reactions that generate these gases involve moles. Problems involving waste handling are also appropriate because they can teach students the importance of understanding concentration units in aqueous solutions. Here is an example of a series of questions that are found in *The Extraordinary Chemistry of Ordinary Things, 2nd Ed.*, by Carl. H. Synder (3):

> According to the pollution limits for Dade County, Florida (shown on page 370, Table 13.6), could 10 liters of solution that contains a total of 0.6 milligram of arsenic be discharged legally as industrial waste water? Could this solution be discharged legally into sanitary sewers? Does this solution meet the drinking water standard for this region? How much water (if any at all) that is uncontaminated by arsenic would you have to add to the 10 liters of solution in order to meet the arsenic standard for each of the categories mentioned above? Consult the MSDS for arsenic and describe the potential hazards associated with high levels of arsenic in water.

This problem requires students to perform fundamental calculations of concentration, convert units from grams per liter to ppm, read and analyze a table of data, draw inferences from their calculations, and also to make decisions about their results. Such problems are important in the education of our students, because we want to foster this type of analytical and critical thinking towards "real world" problems.

Research Labs. Undergraduates who are required or elect a research project with a faculty member must undergo the same safety training and orientation that applies to graduate students, post-doctoral associates, and any other visiting scientist to our laboratories. The chemistry department has enacted various safety training programs initiated by the Ames Laboratory to assist with the safety education of new researchers. Such training involves an initial three-hour orientation followed by specialized videos that address certain aspects of laboratory and equipment protocol.

The Graduate Program. Major efforts have initially taken place within our graduate program, which involves approximately 150 graduate students annually. These students are active not only in chemical research, but also as teaching assistants in our laboratory and some lecture courses, and will greatly influence the attitudes and practices of sound chemical hygiene among our undergraduates. Furthermore, our graduates take leading roles in industrial and academic research environments. During the past three years, we have begun a course for new graduate students in chemistry to review proper chemical hygiene, and last year we began a student safety committee as part of their continuing education, which is designed to treat enforcement and administrative aspects of the responsibilities associated with safety.

Role of the Teaching Assistant. Teaching assistants in general and organic chemistry laboratory courses usually perform their duties in two three-hour sessions every week. Since there is no formal lecture to accompany the laboratory course, instruction comes completely from these teaching assistants. Therefore, our teaching assistants impart tremendous influences on our undergraduate students regarding overall chemical hygiene. This influence involves (1) establishing prudent and proper practices in the laboratory; (2) demonstrating appropriate attitudes towards chemical hygiene; and (3) enforcing course, departmental, and university regulations for conduct in the laboratory. Prudent practices at the general chemistry level mean performing sufficient and necessary pre-lab preparations, maintaining an informative and complete notebook, becoming familiar with new equipment, and learning how to properly dispose of chemical waste. These issues can be directly linked to the laboratory grading policy. Demonstrating appropriate attitudes towards chemical hygiene, however, is intangible. The most effective teachers will emphasize that "chemistry is fun," while managing to develop a student's respect (and not fear!) for chemicals. In order to achieve this among both domestic and international students, we provide a two-week course in "Teaching Methodology," which includes practice in classroom presentations, basics of chemical and fire safety training, as well as instruction in various pedagogical techniques. This course is a fundamental part of the orientation of new graduate students to the chemistry department.

Chemistry 550. This course is offered as a one credit course (one contact hour per week) during the Fall semester for all new graduate students in our chemistry department. Chemistry 550 is a general laboratory safety course designed for chemists working or teaching in a chemical laboratory or related facility. Performance is based upon attendance and brief in-class quizzes. The course utilizes material presented in *Prudent Practices in the Laboratory, Handling and Disposal of Chemicals*, which was published by the National Academy Press in 1995.[1] Table II summarizes the list of topics covered in this course. Presentations are given by various chemistry faculty and staff from the EH&S department. The course involves ACS videos, reviews the ISU Chemical Hygiene Plan, and discusses various topics such as chemical storage, waste management, radiation safety, gas and cryogenic use, and vacuum technology. Upon joining a research group, in-depth training is offered for the specific instrumentation and procedures utilized in that group: Documentation is maintained by the graduate student's research advisor.

Table II. Schedule of Lectures in Chemistry 550

Lecture #	Topic
1	ACS Video: *Introduction to Laboratory Safety*
2	ACS Video: *Protection Against the Odds*; Engineering controls, including fume hoods, glove boxes, and shielding
3	ACS Video: *Safe Laboratory Procedures*; Emergency procedures
4	ACS Video: *Chemical Safety and Environmental Regulations*; OSHA,[5] EPA, and DOT regulations
5	ISU Chemical Hygiene Plan:[6] Managing Risks
6	Compressed Gases
7	Chemical Storage Issues
8	Chemical Waste and Disposal
9	Radiation Safety
10	Radiation Safety (cont.)
11	Laser Safety
12	Low and High Pressure Work
13	Review

Chemistry Safety Representatives. In order to ameliorate the enforcement and review of regulatory compliances in our research and teaching laboratories, we have formed a committee of third and fourth-year graduate students (one from each research group) called the *Chemistry Safety Representatives*. This committee of students conducts biweekly meetings with staff members from EH&S to review aspects of proper chemical hygiene and safety management. Issues such as chemical storage (compatibility, labeling, dating), fire safety (electrical safety), hood operation and management, waste disposal (EPA regulations and recommendations), as well as documentation (inventories, training records, standard operating procedures) are reviewed. Then, as an additional part of each meeting, this committee visits two research or teaching laboratories to "perform an inspection." The students are expected to make and record observations, which are reviewed immediately by EH&S staff members who accompanied them. Their observations are filed with the department and the professor in charge of the laboratory. During the first year of this exercise, we have found this method of safety enforcement to be exceedingly effective in reducing many typical problems: these involve proper labeling of chemical reagents, in particular those synthesized in the lab, hood usage, waste storage, storing chemicals compatibly, and maintaining current training records.

Future Developments

Since Chemistry 550 was incorporated into the graduate curriculum three years ago, every graduate student in the chemistry department at Iowa State University has now received and will continue to receive this fundamental training offered by the department. The committee of chemistry safety representatives noticeably increased the overall department's awareness and attention to several chemical safety issues, and thus represents a type of continuing education for our graduate students. Future efforts for safety education for our graduate students are designed at maintaining the continuity of these two programs.

Our next major thrust, however, is towards the undergraduate curriculum, in which we must address issues for chemistry majors as well as nonmajors. In Fall, 1997, we shall implement a *Proper Chemical Hygiene and Safety* component in all introductory chemistry laboratories, which is designed to be generalized and then incorporated into our advanced laboratory courses. The plan was created because of the review of our usage of *Working Safely with Chemicals*, C. Gorman, Editor and published by Genium Publishers (2). Students recommended a more concise, yet equally informative pamphlet on proper chemical hygiene. Unfortunately, *Safety in Academic Chemistry Laboratories*, published by the American Chemical Society (4) provides too much information for the first-year student in our courses. Therefore, we are creating our own set of instructional materials. Table III lists our proposed "syllabus" for this academic year. This table lists topics and some corresponding exercises for the students, which can be used either as a pre-lab problem, or as part of the final laboratory report. Some of the topics span several weeks to give students time to study and work with the fundamental concepts of these topics. As we work with this material, our goal is to incorporate these ideas into the advanced chemistry laboratories, but to design exercises which are specific to the concepts addressed in the specific laboratory course.

Additional future efforts, which are still in the discussion and planning stages, may include:

- offering or requiring Chemistry 550 or a similar variant (as a short course) to all undergraduate majors passing through the chemistry department curriculum. One possibility is to offer an interactive safety training course which would have to be completed by students before completing their chemistry lecture/lab course. The program would record time spent on programs and would offer quizzes on each subject covered. Students could work at their own speed outside the formal classroom, but provisions for questions and answers would be made during the lecture or laboratory periods.

- assigning a faculty member who is responsible for the safety element of the department to the Undergraduate Affairs and Curriculum Committee or other committees responsible for course content.

Table III. Proposed Syllabus for Proper Chemical Hygiene and Safety as part of ISU's Undergraduate Chemistry Laboratory Courses

Topics	Exercises
1. Personal Protection & Lab Protocol (1 week)	Diagram of Lab
2. Safety Equipment & Emergency Procedures (1 week)	
3. Terminology: Chemical Terms (flammable, combustible, etc.); Exposure Limits; NFPA Signs and other Symbols (3 weeks)	Simple Calculations; Find and Interpret NFPA Signs in Chemistry Building
4. Material Safety Data Sheets: Information; Who uses them; Evaluation (4 weeks)	Find MSDS for chemical in lab experiment using World-Wide-Web; Condense and Evaluate the Information
5. Federal Agencies: OSHA, EPA (2 weeks)	

- having undergraduate students assist in the design and writing of standard operating procedures for their laboratory projects.

- establishing a video library such as the ACS series that can be viewed at a student's convenience. Some sort of documentation of viewing, such as completion of questions from the video subject could be required.

Acknowledgments

The authors wish to thank Steve Heideman (the Chemical Hygiene Officer for the chemistry department), L. Keith Woo and John D. Corbett (instructors for Chemistry 550), George Kraus (chair of the chemistry department), David Inyang, Louis Mitchell and Emery Sobottka (EH&S), and the Ames Laboratory for their assistance and contributions to these efforts.

Literature Cited

1. *Prudent Practices in the Laboratory*; National Research Council, National Academy Press; Washington, D.C., 1995.
2. *Working Safely with Chemicals*; Gorman, C., Ed.; Genium Publishing Corporation; Schenectady, New York, 1994, 2nd Edition.
3. *The Extraordinary Chemistry of Ordinary Things*; Snyder, C. H.; John Wiley & Sons, Inc., New York, 1995, 2nd Edition, p.384.
4. *Safety in Academic Chemistry Laboratories*, American Chemical Society, Washington, D.C., 6th Ed., 1995.
5. OSHA 1910.1450—Occupational Exposure to Hazardous Chemicals in Laboratories.
6. Iowa State University Chemical Hygiene Plan, Second Revision, April, 1997. See also http://www.ehs.iastate.edu/msds.htm for links to this plan and other related areas of the ISU EH&S department.

TECHNIQUES FOR ACHIEVING
AN OSHA SAVVY CHEMIST

Chapter 11

Evaluation and Refocusing Chemical Safety Practices and Instruction at a Mid-Size College

E. Eugene Gooch[1], Eugene Grimley[1], and Paul Utterback[2]

[1]Department of Chemistry, Elon College, Elon College, NC 27244
[2]Department of Safety and Environmental Health, Oregon Institute of Technology, Klamath Falls, OR 97601

Chemical safety education and planning is an important focus of the Elon College experience. An explanation of why safety is important within the Chemistry Department and some of the techniques are provided. Because education is the primary goal, links between safety programs and student development are explored. The Chemistry Department plays a strong role in overall campus chemical safety. This significant contribution requires a consistent departmental self-evaluation and a strong public image.

One scenario of a departmental chemical safety program is presented here. The "Laboratory Standard" (*1*) and the Hazard Communication Standard (*2*) are integrated in order to serve both the Chemistry Department and the overall campus. A historical summary of nearly a decade of work is found in Table 1. The overall philosophy is to use standards as a minimum guideline and as a tool when designing a complete, comprehensive safety program reaching into the Chemistry curriculum. The issue of "compliance" is purposely made non-threatening by focusing beyond minimum guidelines to safe handling practice. Compliance is primarily used as a tool during administrative negotiations to improve safety protocols, equipment and supplies.

Chemical safety planning is an important focus of the overall campus and community. As a result, the act of providing a complete chemical education includes the use of published chemical safety regulations. The intent of these standards is emphasized. The goal is to achieve safe chemical handling behavior in faculty, staff, students and the community at large. Committed time and economics are sometimes a burden, so effective inexpensive solutions are a priority. A historical perspective, some specific solutions, and general safety recommendations are provided here from our "mid-size school" perspective.

Table I. Historical Sequence of Safety Improvements

1. Team Assessment	4. Stock NFPA Labeling
Site	5. Stock Classification
Students	6. Chemical Safety Training
Custodians	7. MSDSs Completion
Instructors	8. Safety References
2. Chemical Stock	9. Chemical Hygiene Plan
Inventory	10. Chemical Hygiene Officer
MSDSs	11. Personal Protective Equipment
3. Hazardous Waste	12. Disaster Preparedness
Segregation	13. Student Awareness
Removal	14. Policy Enforcement

Elon College History and Background.

Established in 1889 in North Carolina's Central Piedmont, Elon College is a comprehensive Liberal Arts college conferring the B.A., B.S., M.B.A, M.Ed., and M.P.T. degrees. There are currently 40 Science Fellows out of a total student body of 3500, primarily from the Eastern Seaboard. A North Carolina Teaching Fellows Honors Program provides science education methods courses, including classroom safety. A strong Honors Program, Isabella Cannon Leadership Program and Elon Science Fellows Program have links to science research and education.

The North Carolina Department of Occupational Safety and Health (DOSH) provides a series of informational packets and consultative advice to those pursuing chemical safety. Although the primary focus is on industry, several of the larger schools have received some enforcement inspections by DOSH. The primary regulatory concerns for Elon College stem from fire protection (highly publicized plant fire in Hamlet, N.C. in 1991) and an incident of death resulting from inappropriate mixing of household chemicals in Dobson, N.C. These well-publicized issues, and others, are motivational factors in overall chemical safety improvements.

Alamance County, N.C. won an award for it's Local Emergency Planning Committee (LEPC) in 1994. The applied Emergency Preparedness Program included planning for local chemical manufacturers, a science supply house and Elon College. The campus has an elaborate Disaster Plan addressing a local rail line, hurricane history and flood history. Additionally, the campus has a strong Safety Committee, Confined Spaces Program, Hazard Communication Program, Bloodborne Pathogens Program and comprehensive safety training events. The strength of these programs is augmented by active participation from the Chemistry Department. Chemistry Department improvements are frequently linked to benefits applied to other campus safety programs.

Discussion (1987-1997): Why Safety Education is Important.

The first year assessment (1987) included assessing problems and planning future goals. The department needed cleaning, better safety practice and a clear commitment to renovations. The current faculty needed to present a unified, strong image in order to achieve departmental upgrades, and a proposed new building. The idea that "safety sells" became very important. A department with a healthy, strong, safe image goes a long way toward competing for, and receiving needed renovation money. In this regard the overall safety plan needed to go beyond compliance, and be perceived as a serious component of everyday practice. Faculty, staff and students all needed to be included and focused on the overall goal of moving forward. Figure 1 summarizes the priority of events leading to successful advancement of safety within the Elon College Chemistry Department.

A list of basic philosophies that drive the need for overall safety improvements was essential for planning (Table II.). Note that compliance to 1910.1450 was only a part of the overall scenario on this list. The assumption here is that regulatory compliance is easily achieved when other concerns are addressed. The larger, overall issue of "doing the right thing" drove the evolution of safety within the Chemistry Department. A hidden benefit to focusing away from compliance as a primary target is that arguments about compliance were minimized. Instead of arguing and focusing on what was bad, the goal was to always move forward building on what was positive.

Table II. Reasons for Campus Health and Safety.

1. Chemistry Department service to the campus and community as chemical experts.
2. Commitment to advanced training for all members of campus.
3. Commitment to best education practices.
4. Commitment to successful laboratory management.
5. College liability for all facilities.
6. College liability for waste management practice.
7. College liability for student health and wellness.
8. College liability for employee safety and workman's compensation.
9. College liability for Faculty Actions.
10. Compliance with published safety regulations (OSHA).
11. Compliance with published hazardous waste regulations (EPA).

First Year Assessment and Challenge (1987).

Staffing was minimal when a decision was made to strengthen the Chemistry program. A high turnover rate in custodial staff made planning and necessary chemical safety training very hard to accomplish. Duke science building was 70 years old in 1987 and lacked proper features such as restrooms, elevators and emergency escapes from all laboratories. Faculty and most departmental offices were only accessible through laboratories. The laboratories were not planned for modern Chemistry and required frequent renovations. Chemical stock was not centralized, properly stored or ventilated. Cabinets, refrigeration, cleaning areas and fume hoods were all in need of upgrades. Unused, dangerous chemicals were accumulated without planning; a major stock reduction was clearly needed. The available stockroom was fully integrated with equipment, instrumentation and chemicals, leading to unnecessary employee interaction with chemicals when accessing other stock. Safety policies were not evident. Safety education was not a component of chemical education. Administrative commitment to chemical science and proper chemical education needed improvement. The idea of achieving the goals and intent described in 29 CFR 1910.1450 seemed unattainable. Clearly, safety was not going to occur overnight and an overall planned commitment was necessary.

Team Building as a Solution.

One of the first decisions was to increase safety awareness by hiring a staff that would participate in the development and implementation of a comprehensive safety program. This eventually included an advertisement specifically for a Chemical Hygiene Officer in 1992. As new staff and the CHO came aboard upgrade issues were addressed. An overall effort to expand the image of the department on and off campus was pursued. Staff members were involved with campus and community planning, contribution to local and national organizations, and integration with other services and academic departments. Where necessary, staff received training, including the Lab Safety Workshop (3). The laboratory plan was eventually used as a template for campus Hazard Communication planning. Safety manuals and policies for students actually became templates for other staff on campus. This validated safety for the students and presented a unified campus program. Films, training packets and handouts were developed for use in both the laboratory and general hazard communication. Student projects from both the sciences and non-sciences focused on chemical safety in occupational and environmental settings. The integrated, team approach addressed many chemical safety questions and legitimized the use of safety in the chemistry curriculum.

Expected Compliance Resulting from Pre-Planning.

Effective January 31, 1991 all laboratory employers were expected to comply with the "Occupational Exposure to Hazardous Chemicals in Laboratories" (the "Lab Standard") ruling 29 CFR 1910.1450 (*1*). Prior to that date, the Hazard Communication Standard 29 CFR 1910.1200 ("Hazcom") was the primary source of chemical safety regulation (*2*). The Lab Standard requires developing a Chemical Hygiene Plan and naming a Chemical Hygiene Officer, and Hazcom requires developing a Hazard Communication Program. A small campus is often forced to look for links that effectively accomplish the goals of both Standards while minimizing time and financial commitments.

Questions of compliance to the 1910.1450 "Laboratory Standard" arise for smaller colleges and universities where science instruction routinely includes laboratory work. What does compliance mean in this situation? Are students and teachers targeted by this standard? Should a "small school" be concerned with compliance? How is compliance achieved and maintained? How does the overall campus Hazard Communication Program connect with a smaller Chemical Hygiene Plan?

Combining the Hazard Communication and Laboratory Standard at a smaller school leads to the following list of basic necessities:

1. Placing someone in charge. A "Chemical Hygiene Officer" who may also serve as a Campus Chemical Safety Officer.

2. Designing and maintaining written plans and inspections for both Standards.
   - Chemical Hygiene Plan (CHP)
   - Hazard Communication Plan (HCP)
3. Inventory of all chemicals campus-wide.
4. Assessment of all chemical hazards campus wide.
5. Collection and maintenance of campus Material Safety Data Sheets (MSDS).
6. Development and enforcement a successful labeling system.
7. Establishing communication mechanisms including employee training, emergency numbers, radios, signs and warnings.
8. Maintenance of appropriate injury logs and incident reports.
9. Effective spill procedures and emergency protocols.
10. Performing hazard assessments, personal protective equipment assessments and safety inspections.
11. Hazardous waste assessment, manifesting and record keeping.
12. Scrutiny of all chemical handling procedures including shipping, receiving, storage, use, and waste tracking.

Whom does the Law Address?

This is a common question. The answer is simple or complicated depending on how the question is approached. The OSHA Standards, including the Laboratory Standard exist for those covered by Workman's Compensation. This usually means anyone receiving a paycheck for the academic institution, including student employees, work-study students, research assistants, teaching assistants, adjunct faculty, visiting scholars, staff and faculty. In some states the OSHA standards apply to classroom and laboratory students specifically. All OSHA standards are written and developed by professionals, so it is advised that they be used with a high level of integrity. In addition to OSHA standards there are waste management regulations enforced by various EPA branches. The overall chemical management plan effect can effect the overall campus image and liability. It is often easiest to treat all campus members as covered by all OSHA and EPA regulations. The overall goal is to eliminate risk and become good stewards of chemical use. Saving money though stock reductions, waste fee elimination, compliance cost avoidance, elimination of expensive hospitalization and experience modifier insurance rates are added benefits.

Identifying Chemical Hygiene Officer (CHO) Duty.

It is the duty of an assigned Chemical Hygiene Officer (CHO) to establish a working Safety Program that meets all Standards, but, more importantly, the CHO must create a comfortable and responsible program that all employees and students will actually use. Simply complying with Standards is not enough. Employees must not only **BE** safe, they must also **FEEL** safe. Employees must buy into the program and play an active role in their own safety. It is not an easy task to establish a safety program of this nature, so the CHO cannot simply be a casually assigned duty. Some institutions have chosen to interpret OSHA's Laboratory Standard very literally and actually hire an independent CHO. At minimum an internal designation must be made. A new employee may be more effective at enforcing safety policies that commonly receive poor acceptance. The title "Chemical Hygiene Officer" can be misleading. It is inevitable that safety duties will expand beyond hygiene to encompass overall safety (non-chemical Standards), education, and waste management. Recommended CHO duty lists and training requirements can be found in Chemical Health and Safety articles (4). The college setting requires the additional duty of interfacing with students and professors who often consider themselves outside the scope of OSHA practice. The considerations listed in Figure 3 may help delineate the role of the CHO in academic policy, student behavior, professorial behavior and the overall academic experience. The focus is on integrating safety with chemical education.

Often, there are instances where the CHO is pursuing safety improvements that appear to be a financial strain on the institution. Be prepared for this potential problem. All parties will need to know the OSHA Standards and address them as responsible minimums. Avoid using threatening language such as "it's the law, so do it". This language can lead to resentment and arguments about regulatory interpretation. Remain firm in commitments to safety by focusing on accepted health and safety practice. Solving inexpensive problems of communication, education, awareness, written plans, documented protocols, MSDSs and inventory control often facilitates renovation requests. The CHO has to establish clout in order to argue for costly improvements, and this clout is established by successful program development. The necessity for improvements can be a gray area. Many of the large ticket items such as ventilation, room temperature control, fire suppression systems, storage cabinets and etc. cannot be solved without spending money, but here many be suitable alternatives. Having a strong safety program with joint participation usually helps sell necessary renovations.

Many Standards are *performance-based* and do not provide *criterion-based* details. Be wary of this difference when fighting for facility upgrades. Be sure to use the published Standards only as far as they apply. Focusing beyond Standard compliance, while referencing the Standard as a tool, is often a more fruitful approach. Commitment to future growth and objective use of published information demonstrates a true concern for safety. This sincerity is more apt to achieve administrative commitment in an educational institution than compliance to regulations.

Table III. Issues to be Clarified for CHO Position.

Extent of enforcement authority, and in which situations.
Extent of authorship on written programs and policies.
Developing recognized, consensus from faculty.
Developing real consequences for policy violators.
Inclusion in the seminar process.
Inclusion in publication efforts.

Chemical Stockroom Reclassification and Shelving

Stock management is one of the first and largest commitments made by a Chemistry Department and the CHO. Commitment to chemical stock management means time, money and space allocation. A large portion of the overall safety program is directly linked to stock management. This includes training, inventory, MSDSs, labeling and PPE policies. An authorized individual or group needs to have ultimate control in order to address the variety of inter-related issues (Table IV). A shared, open stockroom needs a clear set of rules and overall supervision. Maintenance records of storage location, inventory, MSDS availability, labeling requirements and addition/deletion dates are essential. These records can be housed with the safety plan in an open-access area.

A set of rules clarifying and governing security and access are often necessary. The stockroom should be locked when unused. The number of users should be limited so that training can be consistent and focused on specific storage, handling and labeling protocol. Chemical splash goggles should be worn at all times when entering the stockroom. Make every effort to house stock below six feet as part of accident prevention and disaster preparedness. Isolated ventilation (4-6 room changes per hour), temperature control and fire protection systems all must be considered.

There are many potential cross-reactions inherent in alphabetized stocks, especially when organics and inorganics are not separated. One MINIMUM storage schematic is to separate stock into the classes of health hazards, flammables, oxidizers and corrosives. Acid/base cabinets, solvent cabinets and spark-free refrigerators are usually necessary. Further class separations can include acids, acid salts, alcohols, amines, amides, bases, biochemicals, carbohydrates, dyes, esters, hydrocarbons, halogenated, inorganic salts, ketones, metals, nitrogenated, organics, phenols, solvents and surfactants. Alphabetical arrangement within these classes is generally safe, but check to be sure. It is further recommended that nitric acid be stored alone, picric acid be deleted from stock, water-sensitives be stored separately under nitrogen or kerosene and extreme health hazards kept locked. Free classification and storage recommendations can be found on the MSDS, in the Flinn Reference Manual (5), OSHA manuals, NFPA guidebooks (6) and the Prudent Practices Series (7). The Laboratory Safety Workshop (3) also has an extensive loan library.

There is an important educational link to stock handling. Whereas labels and MSDSs are often viewed as the cornerstone to general safety, chemical stockroom management is the more likely cornerstone in an academic setting. Chemical stock is handled every day and leads to overall program development through policy and facility renovation. A potential problem associated with stock is the accumulation of "ready" chemicals for the curriculum. It is tempting to maintain this stock within the classroom grouped by specific exercise needs, but this is can be dangerous and unhealthy. On the other hand, if appropriate storage cabinets and space are limited, it may be wise to avoid common storage of all classes in one stock area, but assess this situation carefully. An unventilated area with no climate control or fire system is not a good place to maintain all solvents, oxidizers, corrosives, and organics. The use of MSDSs, labels and PPE also is significant in this situation. Storage schematics should be reviewed regularly by all users, including students who are potentially exposed. The issue of stock location and access requires that attentive application of chemical safety knowledge be practiced by all.

Table IV. General Stockroom Considerations

Place glassware and other apparatus in center of storage room.
Assessment of potential corrosion of lights, fans and circuit boxes.
Assessment of potential corrosion on delicate instrumentation.
Chemicals around periphery, limited exposure when accessing equipment.
Highly toxic chemicals and precious elements secured.
Custodial chemicals and supplies separated from stockroom supplies.
Isolated hazardous waste from other products.
Stock areas ventilated by isolated system.
No stocks in fume hoods.
Spill preparedness supplies clearly marked and appropriate for stock.
Explosives, ethers, and peroxide-former refrigeration.
Only small-scale storage allowed in laboratory rooms
Access to main stockroom restricted.
Strong inorganic acids and bases stored on bottom shelf.
Biochemicals isolated.
Indicators isolated, level 4 biohazards under lock and key.
Ion Exchange Resins isolated.
Inorganics separated form organics.
Organics separated by class
Solvents separated in solvent cabinets
Oversized bottles on bottom shelf
Only low hazard classes stored 6 feet or more above the floor.
Shelving plastic installed.
Place Inventory on tape for full transcription.
Place inventory on computer for referencing.

Hazardous Waste

A significant portion of stockroom management and CHO duties includes the assessment and proper handling of hazardous waste. The safety program interfaces with waste management practice in that overall awareness is needed by all chemical handlers in order to achieve proper management. This includes avoiding careless sink disposal of student-generated wastes that are otherwise regulated. Many users are more apt to watch their own behavior in the name of environmental protection rather than occupational safety. The overall safety program can benefit from this inherent attention. One quick link comes with proper labeling and MSDS use for waste disposal methods.

When evaluating laboratory waste streams consider recycling, sharps, glassware, blood contaminates, microbial contaminates, preserved specimens, organic tissues, radiation sources as well as chemicals. Each stream has it's own characteristics that contribute to program development. Chemical waste is commonly broken into metals, combustibles, corrosives, reactives, solvents, halogenated solvents, inorganic salts oxidizers, gases, cylinders, mercury, mercury compounds, cyanides, arsenics. Be aware of local regulations and the packaging requirements of projected removal companies. Do not ignore the cross-reactivity storage rules by co-mingling all forms of waste in one designated area. Consider designating small areas in the general stock to house waste by the categories already delineated. If you know your waste removal company, have them inspect your potential streams before you generate and separate wastes. Considerable cost savings occur when you generate and package waste according to a planned removal system. Document your satisfaction with the planned waste handling from "cradle to grave."

Minimizing excess stockroom accumulation is often a problem. Typical concerns of "we might need this some day" and "why spend the money, it's not hurting anything" can retard the best stock management practice. There is an inherent liability from housing old stock. Disaster, emergency and exposure pre-planning requires a level of scrutiny. Additionally, time and money is spent managing unused stock during inventory, cleaning and space allocation. Try tracking all chemicals by placing a sticker on each bottle that is used. Change the colors used each year and record the handling results. Record manufacture date on inventories. This provides an opportunity to assess chemical stock using paperwork rather than wandering about the stock area passing viability judgement. Focus on minimizing specific chemicals by assessing the quantities needed. A removal program spread out over a few years will minimize costs.

Labeling

Successful stock management specifically requires proper chemical labeling. A functional labeling system is in turn critically important to the entire safety program. A few selection parameters should be considered during selection (Table V.). Because the label is very public it can be viewed as the cornerstone to safety awareness. The high visibility requires that the label be used consistently and properly. Labeling assessments and application require daily use of MSDSs, standards and safety references. Each chemical needs to be labeled with a recognizable warning system. This necessitates a training procedure for label recognition, including student handlers. Students in the laboratory should have regular test or lab questions that review their label recognition ability. It is recommended that labeling requirements be inventoried away from the chemical stock itself. This allows the user to access and manufacture labels without entering the stock area.

Table V. Important Issues When Selecting a Label.

- Access to comprehensive reference information.
- Visibility on various bottle sizes.
- Ease of understanding.
- Communication of essential information.
- Ease of creating and updating.
- Potential use in the curriculum.
- Amount of training required.

The National Fire Protection Association (NFPA) has established a relatively thorough labeling system for chemicals (6). The "Fire Diamond" label consists of four colors: blue, red, yellow and white indicating health, flammability, reactivity and special concern respectively. The numbers 0, 1, 2, 3, and 4 are placed inside each color to indicate hazard levels of minimal, slight, moderate, serious and extreme, respectively. This label is nearly universally recognized, and labeling information for many chemicals is easy to access. North Carolina requires this labeling system on the outside of all chemical-containing buildings so there is an added link to building management. Some chemicals have not been classified, but can be labeled using the NFPA technical definitions for each hazard rating (6). This is actually a good academic exercise because of the related terminology and use of f.p., b.p. LD50 and etc. The NFPA diamond is only one of several good labeling systems, and is not specifically required by any Standards. It is highly recommended because of the ease of information gathering, high recognition, and acceptance by local fire departments, rapid training, potential educational use, and low cost of installation.

Personal Protective Equipment (PPE) Policies.

Managing stock, hazardous waste and labeling systems quickly points toward the need for personal protective equipment (PPE) polices. Eye protection is usually the first PPE issue addressed. Impact glasses are not chemical splash goggles with indirect vents, and should not replace appropriate splash protection. It takes only one stray drop to harm an eye whether handling large stock or small quantities. Fogging complaints are addressed by using fog-free lenses, so fogging is no longer a legitimate excuse. A policy of stepping outside the laboratory to occasionally clean lenses also helps avoid this issue. Recent reports suggest that contact lenses are safe in laboratory settings (8, 9). Each laboratory should review their policy regarding contacts keeping in mind the ease of policy implementation and enforcement. Some laboratories chose to pursue an aggressive policy because it may be easier to maintain.

Establish an enforceable eye protection policy with real consequences. A hierarchy of one verbal warning, a written warning, loss of lab period/grade, and expulsion from the laboratory course works well at the college level. The policy should be presented in writing the first day of lab, posted in the department and issued at the time of each violation. Keep copies with the Department Chair, and maintain accurate records of all violations. Expect a protest if a student is removed from laboratory. If you are firm and consistent with high expectations, students will respect the concern shown for their safety. A student who is a consistent violator can be made to read the Safety Plan and write a report contrasting it with the OSHA Standards. This helps the plan author as well as the student. The overall commitment from faculty, staff and students is enhanced.

Appropriate clothing should also be addressed by the PPE policy. Governing open-toed shoes, long pants, aprons (rubber preferred), hair tie-backs and gloves can save future confusion. Remember that the policy includes all, including instructors and TAs.

Respiratory protection should be taken very seriously. If respiratory protection is required be sure to consult and train at the level recommended in OSHA Standards. Protective measures mandated under these Standards viewed with the highest level of professional respect and need to be implemented with this in mind. Consult the MSDS or published exposure level limits for best-recommended protection practice (10).

Unified Intensive Instruction.

A smaller college limited in staff and faculty resources needs to address education in all hiring situations. Hiring laboratory staff personnel with Chemical Hygiene Officer experience has a hidden advantage in the field of training. If training requirements can be met within the educational format of a College, then the CHO duty can actually benefit the instructional effort. Use of labeling, MSDSs, references, stock management and policies within the curriculum fits well when "training" is delivered as "education." The CHO provides an educational approach to chemical safety that benefits overall educational practice. The CHO can accomplish this most easily in the laboratory setting, but has a substantial influence on the overall departmental atmosphere. Support for the CHO in this role strengthens the overall chemical education experience.

The CHO can play an active instructional role by performing mock, on-scene testing using "what if" scenarios, by providing seminars that directly link safety to chemistry, by reviewing/grading reports that involve chemical safety, by informing students of developing chemical safety technologies and by participating in laboratory instruction.

Safety Training Strategies for the CHO.

Establishing a learning environment that incorporates safety can be accomplished in many ways. Projects, extra credit points or laboratory exercises focused on interpreting MSDSs and chemical hazards can be appropriate college level challenges. A policy of semester-long individual oral pop quizzes requiring on-the-spot responses to theoretical accident scenarios definitely maintains safety awareness. Conducting fake spills and exposures with colored water also has a dramatic effect! Posting and/or describing recent accident news keeps students involved (consult The Laboratory Safety Workshop and newsletter or the Journal of Chemical Education for ideas). Giving prizes (safety related or otherwise) is an old standby. There are many safety and regulatory internships available around the country. Post safety internships and safety newsletters, stories and updates on a safety bulletin board. Try to present rules and regulations without a heavy hand. Take the time to explain the rationale behind safety rules, but keep in mind that you work for an educational facility. Use established training techniques that are also favorable educational techniques (Table VI).

The CHO may interface with Teaching Assistants (TA) in laboratory safety instruction. Attending preparation meetings provides the CHO opportunity to help identify the safety procedures that a TA may instruct during pre-lab. Guest instruction in pre-lab presentations allows the CHO to become a vital part of chemical education.

Table VI. Training Tips with Educational Value.

1. Keep the question simple, but provide detailed answers.
2. Encourage hands on use of MSDSs, labels, bottles and etc whenever possible.
3. Demonstrate the use of PPE, emergency supplies and labeling.
4. Provide work sheets to students that can be peer reviewed.
5. Use pre and post testing in laboratories only as enforcement and/or refresher.
6. Use humor where appropriate.
7. Ask why a particular accident was viewed as funny when a room laughs.
8. Use pneumonic devices, slogans, catch-all phrases, but explain acronym jargon.
9. Use personal anecdotes, newspaper clippings or WWW stories as examples.
10. Move around the room during delivery, talk to individuals rather than the room.
11. Use cooperative learning, especially when deciphering MSDSs!
12. Include variety, do not focus on one chemical or one MSDS.
13. Use the Standards as a tool, not as "you better or else"!
14. Preview videos and use only the information that is necesssary.

Can We Make Use of MSDSs in the Curriculum?

Using MSDSs in the curriculum is an obvious link between the CHO and instructional staff. There are some advantages to using MSDSs in the curriculum (Table VII.) to illustrate both chemical principles and other, safety-related principles. MSDSs provide a variety of information that links the labeling system, laboratory handling precautions and general safety information. As MSDS content and delivery has improved with time there has been a greater appreciation for use within the curriculum.

Traditionally, MSDSs are merely a housing issue, but they can be used in pre- and post-laboratory quizzes, laboratory mid-term exams, extra-credit questions, lecture exams, report-writing, special projects and as general references.

Table VII. Some Advantages to Using MSDSs in the Curriculum.

Providing complex information in a common delivery format.
Applying MSDS data to the appropriate laboratory label.
Explanation of BP, VP, FP and other fire cautions.
Explanation of TLV, PPM, TWA exposure limits.
Dimensional analysis on information from exposure limits.
Relevance to chemicals actively used in the laboratory.
How chemical science is used to assess safety precautions.

Departmental Safety Ethos.

To achieve a successful safety program in an academic setting students, staff and faculty need to all participate. The overall department atmosphere should convey a sense of safety if a "safety savvy" educational experience is a true goal. A complete chemical safety education should be viewed beyond mere compliance to regulations.

124

Table VIII. Increasing Safety Awareness in the Chemistry Department.

1. Create a safety resource library in a common study area.
2. Include a labeled notebook containing:
   a. The Safety Plan.
   b. The Chemical Hygiene Plan.
   c. The Hazard Communication Plan.
   d. The OSHA Standards.
   e. The chemical inventory.
   f. MSDSs for each chemical.
   g. Labeling requirements for each chemical.
3. Install a safety bulletin board that includes:
   a. Waste Handling.
   b. Goggle Policy.
   c. Labeling System.
   d. Regulatory Updates.
   e. Current Events.
   f. Scholarly Achievement.
   g. Job Posting.
   h. Netscape News.
   i. Campus Homepage.
   j. Understanding Chemical Reactivity Accidents.
   k. Household Chemistry Use and Misuse.
4. "Decorate" with Signs and Posting.
5. Use a highly visible labeling system.
6. Include training sessions as a regular academic event.
7. Emphasize written plan access.
8. Drill emergency response with everyone.
9. Recognize Chemical Hygiene Officer authority.
10. Hire safety-related Work-Study positions.
11. Conduct spontaneous emergency response training.
12. Allow student projects that are safety related.

Using published regulations as basic information in chemical safety education is recommended, but active handling practice requires overall attention to behavior, a process that cannot be regulated. A variety of techniques can be used to enhance the safety experience (Table VIII). Most of these ideas are simple, and provide a recognizable atmosphere of safety commitment. Ultimately, problems with daily enforcement and concern for compliance are eliminated. Overall peer-pressure and matter-of-fact behavior is enhanced, and the department itself spends less time and effort following safety rules.

Annual Check of the Safety Program.

An educational approach to safety needs to achieve specific goals. One of the most important is annual self-review. If the overall program is successful then self-review is facilitated. Some of the key program elements to assess are listed in Table IX. Generally, these issues are all addressed when chemical safety is an active process within the instructional paradigm.

Table IX. Annual Checklist of Safety.

1. Placing someone in charge as Chemical Hygiene Officer.
2. Designing and maintaining a written Chemical Hygiene Plan.
3. Designing and maintaining a written Hazard Communication Plan.
4. Current Inventory of chemical and other laboratory hazards.
5. Obtain and maintain Material Safety Data Sheets.
6. Develop and enforce an active, effective labeling system.
7. Develop and record training efforts.
8. Establish communication: emergency numbers, signs and warnings.
9. Devise and maintain appropriate injury logs and near misses.
10. Maintain spill reports, equipment and emergency protocols.
11. Assess all work processes including shipping, receiving, storage, use waste generation and record keeping.

Advantages of Safety for Future Directions.

Elimination of risk and attention to safety detail help convince administrative powers that chemical science is a profitable and necessary component of higher education. Building renovations are easier to achieve when the department is viewed as a good steward. Chemical safety has impact campus-wide and as such may lead to helping others achieve their goals as well.

References

1. *Occupational Exposure to Hazardous Chemicals in Laboratories*; 29 Code of Federal Regulations 1910.1450; 1990.

2. *Hazard Communication;* 29 Code of Federal regulations 1910.1200; 1987.

3. *Laboratory Safety Workshop.* Curry College, MA. (617) 237-1335.

4. *What Every CHO Must Know;* Chemical Health and Safety; Vol 4, No 3; 1997.

5. *Chemical and Biological Catalog Reference Manual 1997*; Flinn Scientific Incorporated; Batavia, Il; 800-452-1261.

6. 6. *Fire Protection Guide to Hazardous Materials*; National Fire Protection Agency; 11th Edition; Quincy, MA; 1994.

7. *Prudent Practices for Handling Hazardous Chemicals in Laboratories*; National research Council; National Academy Press; Washington, D.C.; 1981.

8. *Contact Lenses and Chemicals*; Segal, Eileen B.; Chemical Health and Safety; Vol 4, No. 3, 33-37.

9. *Contact Lenses in the Workplace: More Than Meets the Eye*; Petreycik, Richard M.; Safety + Health; June 1997; 74-78.

10. *Pocket Guide to Chemical Hazards*; US Department of Health and Human Services; Government Printing Office; DHHS No. 94-116; Stock No. 017-033-00473-1; 1994.

Chapter 12

Creating a Labeling System with Compliance and Educational Value

Paul J. Utterback and Kathy Sale

Department of Chemistry, Oregon Institute of Technology, Klamath Falls, OR 97601

Lack of proper labeling is a common OSHA violation in chemistry laboratories. The intention behind this citation is to point out a failure to communicate good chemical health and safety for each chemical handled in the laboratory. A good label not only describes the chemical in question, but points to long and short term health effects and specific organs at risk from over exposure. Traditionally this detailed information has been communicated through Material Safety Data Sheets that often go unread. The label system used at Oregon Institute of Technology conveys the required OSHA information while leaning on common label systems, specific MSDS information and consistent interaction from the chemical handler.

The Occupational Safety and Health Act (OSHA) specifies that labeling requirements include the proper identification of long-term (chronic) health effects, potential short-term (acute) health hazard, route of entry, and target organ or body system (*1, 2*). Traditionally this "performance requirement" has been met using the "criteria-based" Personal Protective Equipment (PPE) Standard. However, prescribing PPE does not meet the intent of communicating health and safety, as many are discovering through OSHA's inspections process! Inspections are most concerned with employee knowledge and potential unexpected exposures involving uninformed workers (e.g. custodians and other non-laboratory personnel). Seminara and Parsons present a valuable quiz to measure the effectiveness of warning systems, including labels (*3*). Ultimately, the "quiz" points to a need for basic knowledge in order to understand warnings.

Reale states "the need to convey target organs and chronic effects through in-plant labeling is becoming increasingly apparent" in a presentation of effective labels that meet OSHA's intent (4). Clearly poor chemical communication and worker complacency arising from "quick fix" labeling is a very real issue for employers. Also, site-specific considerations are paramount. On-site requirements need to be reflected in appropriate, tailored safety programs, including specific labeling programs. An educational facility is technically in the business of providing knowledge. The overall safety programs at Colleges and Universities should reflect the business approach by presenting systems that can be used within the ongoing educational process.

From OSHA's perspective a good working label is directly linked to safe behavior. When an employee picks up a labeled chemical he/she should receive a signal to slow down, consider chemical handling, and check relevant safety information as necessary. Ensuring that a label is actively used in this manner is difficult. Traditionally, proper label use is included in annual training, but is not always a part of daily handling.

Labeling for Compliance versus Labeling for Education

Container labeling is a critical component of chemical health and safety. An effective label links the chemical world to the worker's world. The presence of a label is essential, but presence alone does not ensure safe chemical handling. Safe work habits can only be accomplished with appropriate chemical knowledge. Labels may even inadvertently encourage complacency as other methods of safety education are overlooked in an effort to apply a "compliant" label. Increased chemical knowledge and safe chemical handling practice should be communicated from the label to the user. Merely installing criterion-based labels, written plans, Material Safety Data Sheets (MSDSs), signs and other compliance requirements does not ensure overall safe behavior. For example, knowing where MSDSs are located is essential, but it is also the employer's responsibility to ensure that employees know how to access and use MSDS information (2). The same knowledge burden applies to labels. Labels must not only meet minimal legal requirements, but are increasingly inspected as an integral part of communicating chemical safety. The labeling system presented here serves as a cornerstone to overall chemical safety education and laboratory safety training. It integrates the use of inventories, MSDSs, written Chemical Hygiene Plans (CHPs), training, Personal Protective Equipment (PPE) and Standard Operating Procedures (SOPs). Safe handling of chemicals is the focus, not mere label application. A clear advantage of this label is its ability to serve as a teaching tool in chemical education.

Defining a Functional Label

A few key concepts should be kept in mind when selecting a labeling system. First, the selected label must be readable at a glance. Quick identification of basic information is usually essential, but it must be balanced by information content. Second, labels should be scalable such that they are not forced to change format or content simply because of size constraints. Third, the label must communicate health and safety information to the handler. Many labels use common communication formats of colors, numbers, and icons. Because many employees recognize these formats they should be strongly considered when developing any personalized system.

Choosing an appropriate labeling system can be difficult. Assess the application, review the existing selections, and consider prices. There are many commercial offerings, so critically review the application. The following review quickly assesses some commercially available systems (Figures 1-5). All these systems are excellent, but keep an eye on what is needed. Some of the more common themes and issues will be highlighted. These highlights should be considered when designing a successful labeling program.

Review of Typical Labels

A common theme in most labeling systems is the use of colors (Figures 1, 2, 5). Traditionally, blue indicates health, red indicates fire, yellow indicates reactivity, and white is reserved for special precautions or PPE requirements. The number scale of concern (0=minimal, 1=slight, 2=moderate, 3=serious, 4=extreme) is also common to most labeling systems. However, take note of common communication discrepancies. Occasionally assigned hazard numbers can fluctuate between 2-4 for a specific chemical depending on which label is used! Using one label system while applying numbers from another label's database is a common error, sending a confusing message to the worker. Decide, up front, how the chosen label will integrate with the overall safety program. If the label uses a combination of rating systems be sure to include that information in training and written programs. However, this potentially confusing practice is not recommended!

The NFPA Diamond.

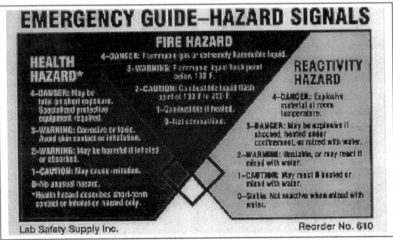

Figure 1. Reproduced with permission from the Lab Safety Supply Catalog (10).

The NFPA label uses a simple number and color code arranged in a diamond. This is a common label, so employee recognition is high (Figure 1). The color-number system is used for quick identification. This fire diamond is designed to inform firefighters of chemical hazards during a live fire. Short-term exposures from fire-fighting events are emphasized in the blue health section. The label does not distinguish between acute or chronic health effects, nor does it specifically indicate target organs, routes of entry, or PPE. However, it does link to a detailed materials guide that further explains chemicals ratings and intent of the system. Many find this guide a valuable component of laboratory safety planning (5).

The HMIS/HMR Label

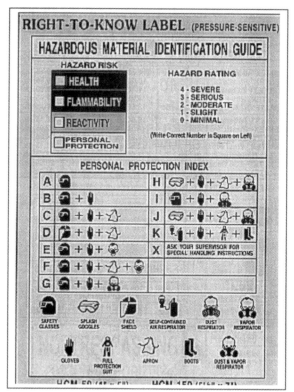

Figure 2. Reproduced with permission from the Lab Safety Supply Catalog (10).

This labeling system uses the common color-number rating system, but the white section is specifically for PPE rating letters (Figure 2). The label can include names of each color designation, or be purchased blank so that the user can clearly write specific chemical information within each color-coded area. The PPE letter rating system must be used in conjunction with key to icons that illustrate specific PPE. Route of entry and target organs are implied from the PPE rating. Acute and chronic health effects are not delineated.

The J.T. BAKER Container/Catalog Label

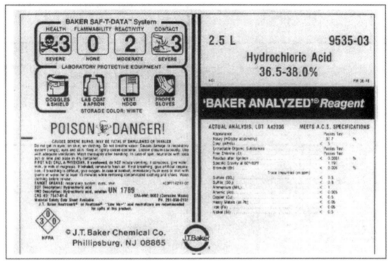

Figure 3. Reproduced with permission from the J.T. Baker Catalog (6).

The J.T. Baker label includes some of the same rating categories as other labels, but the numbering system reflects additional, chronic health hazards (Figure 3). Toxicity is further clarified by an additional category covering skin contact. The remaining target organs and routes of entry are addressed by PPE icons and written warnings. This is a popular rating system in academic laboratories. Written safety information can be found on the label and in the ordering catalog (6).

The Smart Signs Label Using the Genuim Database

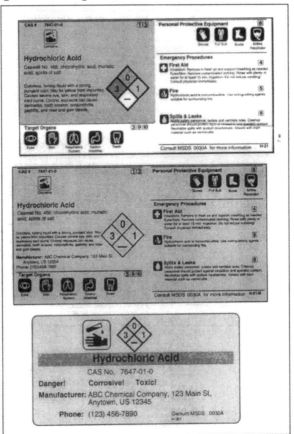

Figure 4. Reproduced with permission from Genium Publishing Catalog (11).

A variety of information and formatting is available from this system. The traditional NFPA diamond is used in conjunction with PPE icons similar to the HMIS/HRS systems (Figure 4). Target organ icons, first aide information, spill procedures and the CAS# are all added features that can be specifically tailored to need. The label can become a "mini MSDS" by including extensive written information. The labels have an additional advantage of being scalable from a database. The user must consider visual confusion. Too much written information can be distracting, leading to ineffective labeling programs.

The "Smart R•T•K" Labeling System

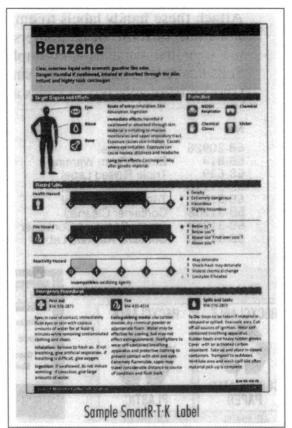

Figure 5. Reproduced from an Advertisement in *Occupational Health and Safety* (3).

These are similar to "Smart Signs", but instead of the NFPA diamond there is a colored "scale" of hazard rating for each category (Figure 5). The scale actually uses the standard 0-4 increment, but delivers the message in a different format. PPE information is expanded to specify what particular body systems are protected, referenced to a human figure icon. The label contains a significant amount of written material and a variety of useful safety information.

Some Typical Issues to Consider

How is good health and safety communicated without undue label clutter? Can a usable label be created that achieves better compliance without having to place an MSDS on the bottle? If the following 3 rules are followed both questions can be answered.

First, legibility is essential. Be aware that too much information is unappealing to the eye, and causes the user to overlook imperative safety information. Second, a successful chemical label needs to be easily manufactured and applied in busy laboratories and diverse chemical stock rooms. Third, the label needs to be applicable to a variety of bottle sizes.

Given these "easy" 1,2,3 parameters the next decision is based on how much information is too much or too little? This decision should be made according to the needs of the facility and its employees. If employee buy-in is low the label receives more burden; training programs should directly focus on day-to-day labels. At OIT, the labeling system has the added burden of being a tool in Hazard Communication training sessions campus-wide. Instead of attending a routine, annual training session, physical plant staff are able to re-designing their own chemical-specific labels each year. Thus trainees create their own, hands-on compliance and demonstrate their actual knowledge as is increasingly recommended in professional safety (2, 7).

An additional consideration is that numbering systems can be confusing because not all numbering systems indicate the same information. For instance, fire ratings are important, but does the rating number indicate fire causing potential or what might occur in an existing fire? Similarly, reactivity presents a myriad of cross-reaction issues in both fire and non-fire situations. Special, reserved sections might deal with corrosives, implicating storage and PPE recommendations, but only indirectly. Many health ratings attempt to describe chemical effects, but health effects are often complicated beyond a simple numbering assignment.

Be careful of forcing a label to cover all issues. What is the chemical variety in the facility? Are there a few, commonly used chemicals or are there hundreds, with many of them rarely used? Is the facility dealing with large quantities or small scale amounts? What are the storage and stock management procedures? What are the waste regulations for the chemical being used? It is tempting to have a label that does it all., but only include what will be covered in training and/or policies.

The Oregon Institute of Technology Label System

This label system is not meant to replace current trends in successful labels. Rather, it utilizes these features in an integrated, educational approach. Other labeling systems have established "norms" such as rapid identification, color coding, number rating, and patterning that should be included in any new labeling system.

It is important to note that the Laboratory Chemical Safety Summaries proscribed in the recent edition of *Prudent Practices* (8) are very helpful is quickly completing the labeling system presented here. The key is to summarize useful chemical safety information relevant to a laboratory worker. It is common for MSDSs to explain chemical safety and handling at a level far above that seen in the laboratory. While the information is practical during a lecture its value is often limited for labeling in the laboratory.

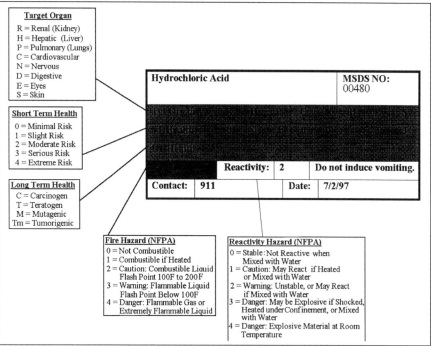

Figure 6. Key to OIT Label Designations

The most significant aspect of the label shown in Figure 6. is the expansion of the common health section (blue) to differentiate between long-term (chronic) and short-term (acute) health hazards. This is perhaps the most valuable "buy-in" aspect for students and workers. Simply pointing out the difference generates a lot of curiosity about chemical behavior (it also results in trainee attention to the details of the lettering/numbering/coloring scheme). Ultimately, the trainees take responsibility for reading MSDSs and developing the appropriate warning systems. Trainees create their own labels, apply their own labels, and actually read MSDSs!

Questions arise regarding atomic, ionic and compound forms of elements and how/why related health hazards arise. In this way the labeling system itself is a good educational tool for chemistry students pursuing health-related professions. The expanded health region provides additional, specific information about the target organ. The enhanced delivery of acute, chronic and target organ information helps explain some of the hidden implications behind specific Personal Protective Equipment (PPE) mandates. The target organ designations also indicate the potential route of entry through specific body systems. An issue that is clarified by this labeling approach is the difference between number ratings for short-term (acute) exposure and long-term (chronic) health hazards as defined by the Hazard Communication Standard 1910.1200. The results are increased worker knowledge, higher levels of safe chemical handling, and, ultimately, an increased level of legal compliance.

Route of entry is covered in two ways. If the health hazard is a "2" or less, the required PPE in the safety policy will protect from exposure. If the health hazard label indicates a "3" or "4" then the safety policy requires referral to the MSDS sheet. The worker then must identify appropriate, specific route of entry, PPE requirements and safe handling behavior precautions.

The Link Between Labeling and Safety Policy

There are a variety of safety polices at Oregon Institute of Technology that link active use of chemical labels. Health hazard warnings and PPE are linked to the label as described above. Standard laboratory behavior, PPE requirements, emergency action requirements, waste handling and other issues are also by specific blank spaces on the label and a reference chart posted in the laboratories and stockroom. While these policies reduce the amount of written information on the label they do not eliminate the need for proper labeling. The label-policy link also provides integrated training information, important feedback on label effectiveness and an increased departmental safety atmosphere.

Precautionary information provided in the overall safety policy further explains common labeling themes. Ultimately, increased safe behavior, enforced by policy, eliminates the need for burdensome labeling information. Some of the Standard Operating Procedures (SOPs) integrated with the label system are listed below.

1. Goggles will not need to be indicated on every chemical label if the Goggle Policy includes proper use at all times when potential exposure may occur. This includes while labs are in session, when research is conducted, and during preparation times. Laboratory pre-lectures and practical exams can be constructed such that there is no potential exposure to chemicals even though the chemistry laboratory is being used.

2. Low to moderate Ingestion Hazard need not be indicated on all bottles if the overall policy states that food and drink are not allowed in the lab. Accidental ingestion from splashing should always be treated and the MSDS should be consulted in the case of this emergency. The implicit assumption here is that students/lab workers will not ingest chemicals intentionally and will respond proactively if exposed to accidental ingestion.

3. Clothing recommendations could be stated in the form of an apron policy. This may be flexible by task, or a requirement for all operations.

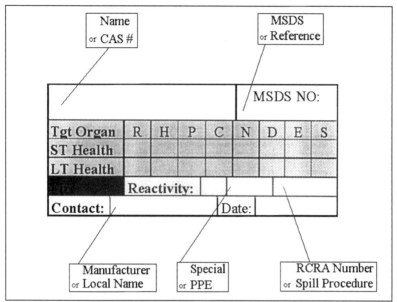

Figure 7. Flexible areas on the OIT label.

Summary of Labeling and Safety Education

The process of using a proper label has become an integrated part of overall safe behavior and is no longer merely an act of compliance to legal standards. The labels that exist on the market are very valuable compliance tools. Be sure to select a labeling system that easily fits in your inventory, CHP and MSDS management programs. The label system presented here is not only flexible and cost-effective, but it also easily fits in existing chemical health and safety programs. If a similar labeling system is adopted by your facility check to be sure that health and safety is properly communicated on a consistent basis.

The bottom line is that the labeling system is only as good as the education that goes along with it. A labeling system should be selected with the facility and the people that will be using it in mind. But the key issue is that people handling the chemicals should be able to tell when there is a potential for a hazardous exposure and how to prevent it! In the end, proper chemical safety education leads to compliance, and, most importantly, safe chemical handling by each user.

138

4. Anything with a Level 3 or higher health hazard requires proper handling to avoid inhalation and/or direct contact. These chemicals may require the use of gloves, a respirator, or a fume hood. The MSDS should always be consulted to determine the proper way to handle these chemicals.

5. Emergency Action and First Aid procedures do not need to be indicated on the label if the policy explains procedures. i.e. eye exposure to chemicals requires 10-15 minute flush using the eyewash available in each laboratory with, once again, referral to MSDS being mandatory. All spills should be immediately reported, many will require evacuation. The supervisor will make this determination.

The label that has been designed for OIT has correlated these ideas to create a labeling system with compliance and educational value. We have many chemicals in our stockroom, a multitude of which are rarely used. It does not make sense to try to train every person that uses the stockroom how to utilize all the chemicals in a safe manner. Instead a person is trained to use the label and the label becomes an ongoing, self-educating process, with the information available when the person is actually going to use the chemical. This proves to be a valuable benefit of the overall system.

Oregon Institute of Technology Label Benefits

These labels are made from a spreadsheet of existing chemical inventory using common software (9). No special purchases are required. A primary benefit of this system is its scalability. Without removing any of the information, the label can be made very small while maintaining legibility. Hazard Communication Standard (1910.1200) terminology allows cross training and inclusion of non-laboratory staff. Particular terminology and definitions can be user specific for various sites (Figure 7). As shown in Figure 7 there are spaces for site-specific information. This allows the user to indicate who prepared the chemical, expiration dates, storage precautions, special cautions, hazardous waste information, spill procedures, and etc. A facility MSDS number is prominently displayed, linking MSDS use and rapid location.

Perhaps the greatest benefit is the ease in which a new, uninformed employee can rapidly adjust to the overall Oregon Institute of Technology (OIT) safety program, simply by using labels. Users are consistently forced to evaluate chemical information during label preparation. In this way the label itself becomes and important education tool!

References

1. *The Hazard Communication Standard.* 29 CFR 1910.1200.
2. *A Complete to OSHA Compliance*; Pederson, Robert D.; Cohen, Joel M.; Lewis Publishers, NY 1996.

3. Seminara, Joseph L.; Parsons, Stuart O. *Occupational Health and Safety.* 1996, June, pp 41-42.

4. Reale, Michael J. *Occupational Health and Safety.* 1996, June pp. 39-40.

5. *Fire Protection Guide to Hazardous Materials*; National Fire Protection Association; 11th Edition, Quincy, MA. 1994.

6. *The J.T. Baker Catalog*; Mallinckrodt Baker Inc.; Phillipsburg, NJ 1997.

7. *The New OSHA: Blueprints for Effective Training and Written Programs;* Daughery, Duane D.; AMA Management Briefing, 1996.

8. *Prudent Practices in the Laboratory: Handling and Disposal of Chemicals;* Board on Chemical Sciences and Technology, Commission on Physical Sciences, Mathematics, and Applications, National Research Council; National Academy Press, Washington, D.C. 1995.

9. *Microsoft Office;* Office Professional and Bookshelf; Microsoft Corporation 1994.

10. *Lab Safety Supply Catalog*; Janeville, WI. 1996. 1-800-356-0783.

11. *Genium Publishing Catalog*; Latham, NY. 1996. 1-518-377-8855.

Chapter 13

Preparation of Experiment-Specific Laboratory Chemical Safety Summaries from Material Safety Data Sheets for Undergraduate Chemistry Courses

David A. Nelson

Department of Chemistry, University of Wyoming, Box 3838 University Station, Laramie, WY 82071

OSHA mandated Material Safety Data Sheets (MSDSs) contain important safety information that should be incorporated into undergraduate chemistry instruction. However, these documents are not well designed for the academic teaching laboratory. Additionally, there is little or no coverage of MSDSs in chemistry laboratory manuals or textbooks. One approach to the incorporation of this topic this is the introduction of a student exercise that involves the preparation of an experiment-specific laboratory chemical safety summary (LCSS) based on the LCSSs presented in the latest edition of *Prudent Practices in the Laboratory* (1). These one-page student-prepared summaries use information obtained from MSDSs, but are modified to the actual quantities, concentrations, and procedures used in the experiment. This approach provides beginning chemistry students with appropriate education about MSDSs.

The "OSHA Savvy" chemist must know how to find, use, and adapt information from Material Safety Data Sheets (MSDSs). Therefore, there should be no argument that education related to MSDSs is appropriate for chemistry students. *Safety in Academic Chemistry Laboratories* (2) states that information from them should be incorporated as part of all laboratory procedures. The National Research Council committee that prepared *Prudent Practices* (1) recommends that students in laboratory classes be included in the safety education appropriate for all other laboratory workers, and that this education, which includes the use of MSDSs, cannot be assumed to be optional. However, an instructor is faced with several problems when considering the introduction of safety related topics in undergraduate chemistry courses. These include the matter of appropriateness, when and where in the curriculum to introduce such material, and accessible sources of information. This Chapter discusses a series of laboratory or lecture-laboratory exercises designed for a general chemistry course which first introduce students to MSDSs, next to the concept of less complex safety summaries, then to the idea of the preparation of an experiment-specific summary along with suggestions for necessary background material and examples of accompanying exercises. The preparation of the experiment-specific laboratory chemical safety summaries (LCSSs) described here was used in a one-semester course in general chemistry at the University of Wyoming (CHEM 1000). The students taking this course come from a variety of disciplines, primarily agriculture and allied health, including nursing. Many of these students will work with chemicals and encounter MSDSs in their professional careers. Since for some this course is the only chemistry course they will take, there is an obligation to present them with the same, if not higher, level of safety education as that received by students in other courses in general chemistry. The safety instruction was reinforced by presenting it in the lecture portion of the course. In CHEM 1000, the first 15-20 minutes of each Monday lecture was devoted to safety issues related to the laboratory of the week. Attendance at this lecture was mandatory, and all students got the same discussion from the faculty member giving the course. Teaching assistants (TAs) were also required to attend the safety part of the lecture. The safety issues were reemphasized at TA meetings, and then in the laboratory sessions.

Introduction to Material Safety Data Sheets

Students in CHEM 1000 were first introduced to the MSDS as a source of safety information in a Safety Handout (5) that is given out in all laboratory classes. This handout covers general University and Department safety policies, and includes a brief description of the nature of a MSDS. This description was adapted from that given in Section III.2.2 of *Safety in Academic Chemistry Laboratories*.

Using the Material Safety Data Sheet as a Source of Information. Although MSDSs are a basic source of health and safety information about chemicals, they are not designed for use in introductory chemistry classes and associated laboratories. Both *Prudent Practices* and *Safety in Academic Chemistry Laboratories* point out several cautions in using them as an educational resource. Students should know that they are OSHA required documents originally mandated by the Hazard Communication Standard 29 CFR 1910.1200 and designed primarily for informing workers who may be involved in handling large quantities of a particular chemical for extended periods of time on a daily basis. OSHA requires certain information to be given, but does not dictate the format. The MSDSs may come from any source, and they are considered as a tertiary source of information. The data presented is not referenced. There is no requirement that they be checked for accuracy by any outside agency. Ref.. 2. lists several typical errors found in MSDSs (3), and ref. 1. details some additional limitations (4).

As long as their limitations are appreciated, MSDSs can be important educational documents if they are used wisely. Since few laboratory manuals (22) and essentially no U.S. general chemistry textbooks even mention MSDSs, instructors must use auxiliary sources of information.

Some consideration was given to the first MSDS for a laboratory chemical that students were actually ask to read in detail. One recommendation is to use an older form MSDS. These are usually shorter, are more narrative, and contain less arcane abbreviations than the American National Standards Institute (ANSI) format. The information on physical properties and precautionary labeling is usually found up front. The first MSDS should represent a chemical that has some obvious hazard already known to most of the students, such as ammonium hydroxide. The point is to introduce the MSDS as a useful document and convey some respect for the information it contains. TAs (and faculty) must be instructed not to belittle or ridicule the MSDS as a useless bureaucratic document. Students should appreciate the fact that a MSDS has to address the worst possible industrial scenario. They should be told that MSDSs for dilute solutions are often the same as those for concentrated solutions or pure chemicals. The same generic statements are often used, in particular the disclaimer that usually states something like "The user should recognize that this product can cause severe injury and even death, especially if improperly handled or the known dangers are not heeded." This statement was on a MSDS for lump iron.

Background Activities. Before attempting to prepare a Laboratory Chemical Safety Summary related to a specific experiment, the students were introduced to MSDSs by a series of preliminary lecture-laboratory discussions. These are outlined below:

- Explain the basic MSDS information specified in 29 CFR 1910.1200 (g) (2) (Table IA)
- Show a typical 29 CFR format MSDS
- Explain that additional information is often presented
- Emphasize:
  A. OSHA regulations require availability in the workplace
  B. Designed to cover "worst-case" scenario in chemical industry
- Show a new ANSI 16 Section Format MSDS

As mentioned above, the CHEM 1000 students were given some information in a general laboratory safety document when they first checked into lab. In addition to the items of information explained there, the 16 Sections of information of the new ANSI MSDS format were covered, emphasizing definitions of the terms used primarily in Section 8 (Exposure Controls, Personal Protection), Section 11 (Toxicological Information), and Section 12 (Ecological Information). The following acronyms found in many newer MSDSs as well as LCSSs, that are not listed in ref. 2., were also explained.

1. NIOSH - National Institute of Occupational Safety and Health
2. IARC - International Agency for Research on Cancer
3. NTP - National Toxicology Program

4. RCRA - Resource Conservation and Recovery Act
5. TSCA - Toxic Substances Control Act
6. CWA - Clean Water Act
7. SARA - Superfund Amendments and Reauthorization Act
8. STEL - Short Term Exposure Limit
9. OEL - Occupational Exposure Limit
10. TWA - Time Weighted Average
11. RTECS# - Registry of Toxic Effects of Chemical Substances Number

MSDS Lecture. At the lecture before the first laboratory using chemicals, appropriate sections of ANSI format MSDSs of the chemicals to be used were projected, and the students were asked to transfer some elementary hazard information from Section 3 onto a form used for recording attendance (Figure 1). A document "Understanding Material Safety Data Sheets" obtained from an Internet page at Stanford University's safety web site (6) was also distributed. Finally, the students were shown how to obtain MSDSs from SIRI, Safety Information Resources on the Internet (7), using a projection computer connected to the Internet. During this lecture the ANSI MSDS Sections were presented again, and compared with those of an older format MSDS. *Safety In Academic Laboratories* is widely used, and often distributed to students and/or teaching assistants. The latest (6th) edition does not specifically discuss the 29 CFR or ANSI format MSDSs. In Section III 2.2 it lists required information from six of the twelve 29 CFR sections. The areas of information neglected are v, vii, viii, ix, x, and xii. (see Table IA).

At the time of this writing the ANSI Format MSDS was not yet required by OSHA, but that requirement should be forthcoming. For that reason, it was instructive to make a listing of the corresponding twelve (i)-(xii) types of information required for MSDSs by the Hazard Communication Standard 29 CFR 1910.1200 (g) (2) with a detailed description of the content of the 16 ANSI Sections, as shown in Table IB. This table was discussed in the MSDS lecture. If lecture time is limited, this could be done as a lab or pre-lab exercise, or given as a handout.

Some of the scope and limitations of MSDSs were discussed. Several copies of the MSDSs were available during the lab period, and the students were given time to look them over. This may be the first time that many of the students have ever seen a MSDS. For all subsequent experiments, the MSDSs were available during the laboratory periods, and were also put on library reserve. The teaching assistants were also given copies, and instructed to have them available for student help sessions. For the next two laboratories, the students were asked to look up the major hazards of the chemicals to be used and submit the data as part of their pre-lab report. The hazards were also reviewed in the weekly lecture safety discussion.

Introduction to Safety Summaries. The concept of safety summaries is well established. Many compilations were available before MSDSs were required by OSHA (8). One of the most obvious is a modern chemical label (See Chapter 12). However, it is a common current practice in the general and organic service teaching laboratories to remove all long-term chemical storage, for a variety of

Table IA. Minimum information specified for MSDS by 29 CFR1910.1200.

Subsection	Information.
(i)	The label identity. (Note: the chemical formula is not required.)
(ii)	Physical and chemical characteristics: e.g. v.p., flash point.
(iii)	The physical hazards: potential for fire, explosion, reactivity.
(iv)	The health hazards: signs of exposure, medical condition aggravated.
(v)	The primary route(s) of entry
(vi)	The OSHA PEL, the ACGIH TLV; also any other used exposure limits.
(vii)	Whether listed as a carcinogen by NPT, IARC, or OSHA.
(viii)	Precautions for safe handling and use: include spills.
(ix)	Control measures: engineering controls, PPE.
(x)	Emergency and first aid.
(xi)	The date of preparation or latest modification.
(xii)	The name, address, tel. No. of preparer or distributor of the MSDS

Table IB. Information Sections of ANSI format MSDS[a]

Section #	Information	Corresponding 29 CFR Subsection
1	Chemical Product and Company Identification MSDS Substance Name, Trade names, Synonyms, CAS and RTECS #, Company ID, Tel#s for More Information, Emergency, CHEMTREC	(i)
2	Composition, Information on Ingredients Chemical Name, CAS#, Percentage	(i)
3	Hazards Identification Emergency Overview Appearance, Odor, Handling Cautions, Label Information, Target Organs Potential Health Effects Inhalation, Eye, Skin Contact, Ingestion, Short and Long Term Effects, Carcinogen Status	(iii)(iv)
4	First Aid Measures Eyes, Skin, Ingestion, Inhalation, Notes to Physician	(x)
5	Fire Fighting Measures General Information, Extinguishing Media Autoignition Temp., Flash Point, LEL, UEL NFPA Rating	not addressed
6	Accidental Release Measures Spill, Leaks, PPE	(viii)

| 7 | Handling and Storage | (viii) |

7 Handling and Storage (viii)

8 Exposure Controls, Personal Protection (iv)(x)
Chemical Name, Exposure Limits for ACGIH,
NIOSH, OSHA final PEL, PPE for: eyes, gloves,
clothing, respirators

9 Physical and Chemical Properties (ii)
Description, mw, molecular formula, bp, fp/mp,
vp, v density, sp gr, water solubility, pH, odor
threshold, evaporation rate, viscosity, solvent
solubility, odor

10 Stability and Reactivity (iii)
General Reactivity, Conditions to Avoid,
Incompatibilities, Hazardous Decomposition
Products, Polymerization

11 Toxicological (Toxicology) Information (iv)(v)(vi)(vii)
Irritation Data, Toxicity Data, LD50/LC50,
Carcinogen Status, Teratogenicity, Neurotoxicity,
Mutagenicity

12 Ecological Information not addressed
Environmental Impact Rating, Aquatic Toxicity,
Degradability, Log Bioconcentration Factor, Log
Octanol/Water Coefficient

13 Disposal Considerations not addressed
EPA Hazardous Waste#, CERCLA Section
RCRA D,F,P,U Series Listing, Reportable Quantity

14 Transportation Information not addressed
DOT Shipping Name, ID#, Hazard Class, Packing
Group, Labeling Requirements

15 Regulatory Information not addressed
Yes/No Status for TSCA, CERCLA Section 103,
SARA Sections 302, 304, 313, OSHA Process Safety,
CA Prop. 65, Sara Hazard Categories for Acute, Chronic,
Fire, Reactivity, Sudden Release, International Exposure Limits

16 Additional Information not required
Modifications to Specific Sections, Creation/Revision Dates

[a] summary of information appearing on MSDSs from MDL Information Systems, Inc. and Fisher Scientific.

CHEM 1000   DATE: 23 September 1996 Attendance Record.

1._____ I certify that I was present during CHEM 1000 lecture on the above date and received the safety information/instruction/exercise that was given related to the laboratory.

2._____ I was **NOT PRESENT** during the CHEM 1000 lecture on the above date when all or part of the safety related information was presented. I realize that I must make up this exercise before I can perform the related laboratory exercise. If I do not make up this safety exercise the lab will be counted as a **missed lab.**

PRINTED NAME _____

SIGNATURE   _____SID#_____

LAB SECTION NUMBER_____ 10 T-7PM, 11 T-8, 12 T-1, 13 W-2, 14 R-2

If 2. is checked:

Safety Exercise Make up Authorized_____

Safety exercise was done: Date_____ Time_____

SIGNATURE of exercise Supervisor _____

STUDENT SIGNATURE_____

Chemical	Major Health Hazard	Major Physical Hazard

Figure 1. Safety Attendance Form.

safety and liability related reasons. All preparation of reagents is done by lab assistants, and beginning students rarely see the actual chemical container and its original label. It is important that beginning students get in the habit of observing the hazard information on labels. In the CHEM 1000 laboratories the actual chemical container was placed in the hood during the lab, or if this was not appropriate the label was reproduced and made available along with the MSDS.

By this time the fact that MSDSs can be rather unwieldy documents had become obvious to the students. We decided to introduce the concept of a purpose-specific chemical safety summary by using the Laboratory Chemical Safety Summaries (LCSS) described in *Prudent Practices* (9). These are two-page documents listing a selected set of data for a single substance which comprise "a concise summary of safety information that should be adequate for most laboratory uses". Their stated use is by trained laboratory workers with a general knowledge in the safe handling of chemicals. In addition to physical properties they address odor, toxicity data, major hazards, flammability and explosivity, reactivity and incompatibility, storage and handling, accidents, and disposal. An example, for Ammonium Hydroxide, is shown in Figure 2. A disclaimer explains their purpose and cautions that they should not be used for non-laboratory applications. It is important to point out that a LCSS is not a substitute for a MSDS.

Lecture-Laboratory Exercise 1: Comparing MSDSs and LCSSs. During the lecture the concept of a LCSS was explained as one was projected (we used ammonium hydroxide for the first example). If an Internet-accessible projection computer is available the 88 LCSSs listed in *Prudent Practices* can be found at the Howard Hughes Medical Institute web site (10). Selected portions of a corresponding MSDS for ammonium hydroxide were also projected. We used a 13 page one from MDL Information Systems, Inc. (11) to emphasize the amount of data reduction involved. The students were told that in the laboratory they would be given a copy of the LCSS and would have to find the information given there on a corresponding MSDS, indicate the MSDS Section the information came from, and compare the data and values given. For this purpose, and for preparation of other LCSSs, the students were given a handout shown as Table II. Four or five MSDSs from different sources comprising ANSI as well as other formats were distributed during the laboratory exercise. This exercise was designed to give the students more experience in using MSDSs as a source of information, and to show them how the data can be selected to be more related to laboratory use.

Laboratory Exercise 2: Preparation of an Experiment-Specific LCSS. A standard MSDS disclaimer statement is "Users should make their own investigations to determine the suitability of the information for their particular purposes." This is essentially the philosophy behind a safety summary. Using the *Prudent Practices* LCSS as an example, and the experience from the previous exercise, during the next lab the students were instructed to prepare a LCSS from MSDS data that applied specifically to the particular experiment they were doing that day. They were told to consider the actual amounts, concentrations, form of the chemicals they would use, the conditions and actual procedures of the experiment,

148

LABORATORY CHEMICAL SAFETY SUMMARY: AMMONIUM HYDROXIDE

Substance	Ammonium hydroxide (Aqua ammonia, ammonia) CAS 1336-21-6
Formula	28 to 30% NH_3 in H_2O
Physical Properties	Colorless liquid bp: unstable above 27.8 °C, mp −71.7 °C Concentrated ammonium hydroxide is a 29% solution of NH_3 in H_2O.
Odor	Strong pungent ammonia odor detectable at 17 ppm
Vapor Density	0.59 for anhydrous NH_3 (air = 1.0)
Vapor Pressure	115 mmHg at 20 °C for 29% solution
Autoignition Temperature	690 °C (for ammonia)

Toxicity Data

LD_{50} oral (rat)	350 mg/kg
PEL (OSHA)	35 ppm (27 mg/m³)
TLV-TWA (ACGIH)	25 ppm (17 mg/m³)
STEL (ACGIH)	35 ppm (27 mg/m³)

Major Hazards Highly corrosive to the eyes, skin, and mucous membranes.

Toxicity Ammonia solutions are extremely corrosive and irritating to the skin, eyes, and mucous membranes. Exposure by inhalation can cause irritation of the nose, throat, and mucous membranes. Exposure to high concentrations of ammonia vapor (above approximately 2500 ppm) is life threatening, causing severe damage to the respiratory tract and resulting in bronchitis, chemical pneumonitis, and pulmonary edema, which can be fatal. Eye contact with ammonia vapor is severely irritating, and exposure of the eyes to ammonium hydroxide can result in serious damage and may cause permanent eye injury and blindness. Skin contact can result in severe irritation and burns; contact with the liquid results in cryogenic burns as well. Ingestion of ammonium hydroxide burns the mouth, throat, and gastrointestinal tract and can lead to severe abdominal pain, nausea, vomiting, and collapse.

Ammonium hydroxide has not been found to be carcinogenic or to show reproductive or developmental toxicity in humans. Chronic exposure to ammonia can cause respiratory irritation and damage.

Flammability and Explosibility Ammonia vapor is slightly flammable (NFPA rating = 1) and ignites only with difficulty. Ammonia forms explosive mixtures with air in the range 16 to 25%. Water, carbon dioxide, or dry chemical extinguishers should be used for ammonia fires.

(continued on facing page)

The information in this LCSS has been compiled by a committee of the National Research Council from literature sources and Material Safety Data Sheets and is believed to be accurate as of July 1994. This summary is intended for use by trained laboratory personnel in conjunction with the NRC report *Prudent Practices in the Laboratory: Handling and Disposal of Chemicals*. This LCSS presents a concise summary of safety information that should be adequate for most laboratory uses of the title substance, but in some cases it may be advisable to consult more comprehensive references. This information should not be used as a guide to the nonlaboratory use of this chemical.

Figure 2. LCSS for Ammonium Hydroxide.

Reactivity and Incompatibility

Highly explosive nitrogen halides will form in reactions with halogens, hypohalites, and similar compounds. Reaction with certain gold, mercury, and silver compounds may form explosive products. Violent reactions can occur with oxidizing agents such as chromium trioxide, hydrogen peroxide, nitric acid, chlorite, chlorate, and bromate salts. Exothermic and violent reactions may occur if concentrated ammonium hydroxide solution is mixed with strong acids, acidic metal and nonmetal halides, and oxyhalides.

Storage and Handling

Ammonium hydroxide should be handled in the laboratory using the "basic prudent practices" described in Chapter 5.C. All work with this substance should be conducted in a fume hood to prevent exposure by inhalation, and splash goggles and impermeable gloves should be worn at all times to prevent eye and skin contact. Containers should be tightly sealed to prevent escape of vapor and should be stored in a cool area separate from halogens, acids, and oxidizers. Containers stored in warm locations may build up dangerous internal pressures of ammonia gas.

Accidents

In the event of skin contact, immediately wash with soap and water and remove contaminated clothing. In case of eye contact, promptly wash with copious amounts of water for 15 min (lifting upper and lower lids occasionally) and obtain medical attention. If ammonium hydroxide is ingested, obtain medical attention immediately. If large amounts of ammonia are inhaled, move the person to fresh air and seek medical attention at once.

In the event of a spill, soak up ammonium hydroxide with a spill pillow or absorbent material, place in an appropriate container, and dispose of properly. Alternatively, flood the spill with water to dilute the ammonia before cleanup. Boric, citric, and similar powdered acids are good granular neutralizing spill cleanup materials. Respiratory protection may be necessary in the event of a large spill or release in a confined area.

Disposal

In some localities, ammonium hydroxide may be disposed of down the drain after appropriate neutralization and dilution. In a fume hood, the concentrated solution should be diluted with water to about 4% concentration in a suitably large container, and neutralized with a nonoxidizing strong acid such as HCl. The resulting solution can be discharged to the sanitary sewer. If neutralization and drain disposal are not permitted, excess ammonium hydroxide and waste material containing this substance should be placed in an appropriate container, clearly labeled, and handled according to your institution's waste disposal guidelines. For more information on disposal procedures, see Chapter 7 of this volume.

The information in this LCSS has been compiled by a committee of the National Research Council from literature sources and Material Safety Data Sheets and is believed to be accurate as of July 1994. This summary is intended for use by trained laboratory personnel in conjunction with the NRC report *Prudent Practices in the Laboratory: Handling and Disposal of Chemicals*. This LCSS presents a concise summary of safety information that should be adequate for most laboratory uses of the title substance, but in some cases it may be advisable to consult more comprehensive references. This information should not be used as a guide to the nonlaboratory use of this chemical.

Figure 2. (cont.) (Reproduced with permission from ref. 1. Copyright 1995 by the National Academy of Sciences.)

Table II. MSDS locations for LCSS Information

LCSS Item	29 CFR type MSDS Subsection	ANSI type MSDS Section #
1. Substance	(i) identity	1.
2. Formula	not required	1. or 9. (not always given)
3. Physical Properties	(ii) Physical and chemical characteristics	9.
4. Odor	not specified	3. and 9.
5. Vapor Pressure	(ii) specified	9.
6. Flash Point	(ii) specified	5.
7. Autoignition	Temperature not specified	5.
8. Toxicity Data	(vi) OSHA PEL, ACGIH TLV, and other exposure limits	8. and 11.
9. Major Hazards	(iv) Health hazards	3.
10. Toxicity	(v) Routes of entry	3.
11. Flammability, Explosivity	(iii) Physical hazards	5.
12. Reactivity and Incompatibility	(ii) Physical/chemical characteristics	10.
13. Storage, Handling	(viii)	7. and 8.
14. Accidents	(viii) implied, not specified	4. and 6.
15. Disposal	not specified	13.

the personal protective devices they had available, and any specific safety procedures they had already been instructed to follow. For CHEM 1000 we designed a one-page "CHEM 1000" LCSS form, shown in Figure 3. This form is less detailed than the *Prudent Practices* format, but was appropriate for a beginning general chemistry class. Data such as Vapor Density, Vapor Pressure, Flash Point, and Autoignition Temperature were omitted   since these were applicable to very few of the chemicals used in this course. For MAJOR HAZARDS the students were asked to pick from the MSDS what they thought were the single main physical hazard and the single main health hazard.   Under TOXICITY they were instructed to decide on a low, moderate, or high rating as implied by the MSDS, whether the toxicity was acute or chronic, list possible route of entry, the LD_{50} or LC_{50}, and whether there were any statements concerning carcinogenicity.

The question related to exceeding the PEL (permissible exposure limit) required the students to do some analytical thinking. Some conditions needed to be presented to, or assumed by, the students. This question was introduced by posing an exercise which assumed that each student had spilled their allotment of some volatile chemical on the bench at the same time, that it had completely evaporated, dispersed uniformly throughout the lab, and that the hoods were not working. This amounted to determining the mg/m^3 concentration. The student needed to make a rough estimate of the volume of the laboratory and determine the total amount of chemical released. This made an interesting problem to test the basic mathematical skills of the students. Having done an exercise like this gave the students a worst-case scenario basis to think about a reasonable answer to the question. A more sophisticated approach to this problem has been presented (12).

The question about hazards and toxicity related to the lab exercise, and how these differ from the MSDS information, represented the crux of the LCSS, and gave students an opportunity to relate some basic chemistry to MSDS information. When students first began to prepare LCSSs, considerable assistance and directed thinking was required. In CHEM 1000 the first time the students were asked to prepare a LCSS related to one of their own procedures was for one of the metals used in an experiment determining density by displacement of water. The metals available were aluminum, chromium, copper, iron (as small nails), lead, magnesium, nickel (as coins), tin, and zinc. Intuitively most students probably would think of most of these materials as innocuous since they are encountered frequently in daily life. Lead may be an exception due to recent publicity about the toxic effects of lead in children. This example focused on the question of carcinogenic hazards. Some of the problems the students had to consider in preparing a LCSS for the nickel metal are as follows. All MSDSs list nickel metal, and any alloy containing nickel as a carcinogen. One of our general departmental safety principles is that carcinogens are not to be used in the large enrollment lower-division laboratory classes, and this policy is written into the material that the students receive. They may well have believed that we were in violation of one of our own safety policies in using this experiment. The fact that the carcinogenic properties of nickel are associated with soluble Nickel II, and not the metallic element can be implied from a careful reading of some MSDSs, but it is often not

152

CHEMISTRY 1000 LABORATORY CHEMICAL SAFETY SUMMARY (LCSS)

NAME_____ LAB SECTION_____

EXPERIMENT #_____ TEXT_____

NAME OF CHEMICAL _____FORMULA_____

PHYSICAL STATE_____CONCENTRATION_____

AMOUNT USED per STUDENT_____

MAJOR HAZARDS FROM MSDS_____

TOXICITY FROM MSDS_____

FLAMMABLE?_____ VOLATILE?_____ODOR?_____

PEL (OSHA) from MSDS_____ .

COULD THE PEL BE EXCEEDED DURING THE LAB EXERCISE? Explain clearly why
or why not._____
_____

HAZARDS AND TOXICITY related to laboratory exercise.
_____
_____

WHAT SHOULD YOU DO IN CASE OF:

 EYE CONTACT?_____

 SKIN CONTACT? _____

 CONTACT WITH CLOTHING?_____

 INHALATION?_____

 INGESTION?_____

 SPILLED CHEMICAL?_____

WHAT IS THE DISPOSAL PROCEDURE?_____

PREPARED BY_____ DATE_____

MSDS REFERENCE_____

Figure 3.   CHEM 1000 LCSS Form.

clear. This fact needed to be pointed out to students. Nevertheless, the issue of the regulatory and legal implications of MSDS information probably should be considered. One MSDS for nickel pellets prepared 12APR88 and reviewed 04MAY 94 (13) listed the following under Ingredients/Identity Information:

Ingredient: NICKEL(SOLUBLE) (SUSPECTED HUMAN CARCINOGEN BY ACGIH, NTP. SUSPECTED ANIMAL CARCINOGEN BY IARC - GROUP 2B)(15)

and under Health Hazard Data:

Carcinogenicity NTP: YES, IARC: YES, OSHA: NO. (15)

Another MSDS for Nickel Metal created on 10/06/1995 and revised on 3/04/1996 (14) provided the following in SECTION 3:

HAZARDS IDENTIFICATION EMERGENCY OVERVIEW: WARNING! CARCINOGEN

and in SECTION 11-Toxicological Information-for Carcinogenicity:(15)
ACGIH: A2-suspected human carcinogen,
NIOSH: occupational carcinogen,
NTP: Suspect carcinogen,
OSHA: Possible select carcinogen (note change),
IARC: Group 2B carcinogen.

The difference in the OSHA carcinogenicity rating was noted. Further reading revealed the evidence for carcinogenicity was from epidemiological studies among nickel refinery workers exposed to nickel fumes and dusts. Although we did eliminate experiments using soluble nickel compounds from our analytical chemistry procedures, we did not remove nickel metal as a density/specific heat sample from our general chemistry experiments. Discussing the relationship between various information given on the MSDSs and the risks to the student in this experiment gave an opportunity to present some of the chemistry of nickel. It was pointed out that no dusts or fumes are generated. Although nickel is above hydrogen in the electromotive series, it is listed as insoluble in water, and stable under normal temperatures and pressures. Thus, it would not be expected to form soluble nickel ions when displacing the water in this experiment. If exposed to acids, however, it would be expected to dissolve and form a carcinogenic solution. As a coinage metal, as a component of stainless steels, and as an alloy component in many other widely used commercial and industrial products persons come in contact with nickel in every day life. These facts might help to explain the position that the hazards in this experiment were not sufficiently high to require special protective procedures when handling nickel metal in the form of coins, or other bulk form. In all cases anyone using nickel metal will have to make their own decision concerning its risk. Students must appreciate that the nickel in coins is the same chemical element that the MSDS is referring to. This is an example of using information from a MSDS to teach some basic chemistry. The OSHA requirement that the MSDS is designed to address workers involved in the production and use of nickel could be made realistic by discussing, in the lecture, the production of nickel from smelting of sulfide ores, and how a possible exposure to nickel dusts, vapors, and soluble nickel compounds could occur.

Similar situations presented themselves when looking at the MSDSs of some the other metals used in the experiment. Chromium is another element with health hazards similar to nickel. A MSDS for chromium metal with a Safety Data Review date of 30NOV92 (16) gave the following information:

> "IARC has determined a 'casual' association between occupational exposures to chromium and certain chromium compounds and cancer in humans. This determination was based on evidence where exposures were essentially to hexavalent chromium compounds. The products covered by this data sheet contain chromium in metal state. The AICGH has concluded that chromium metal is not carcinogenic to humans."

The listings for carcinogenicity by NTP, IARC, and OSHA were all negative. Another MSDS with a Safety Data Review date of 12SEP94 (17) indicated positive carcinogenicity listings by NTP and IARC. The OSHA listing was still negative for the metal, but the fact that OSHA considers hexavalent chromium a carcinogen was indicated. The AICGH statement was not found on this MSDS. A more recent MSDS revised 3/04/1996 (18) listed in Section 11 under Carcinogenicity:

> "Chromium - IARC: Group 3 carcinogen, based on incidences of respiratory cancer in chromium VI workers".

However neither MSDS (17) or (18) made the direct statement that chromium metal was not considered carcinogenic by IARC.

In the case of lead, older MSDSs do not indicate it as a carcinogen. In a more recent MSDS revised 5/28/1996 (19), lead is indicated as classified by IARC as a Group 3 carcinogen (listed as LEAD COMPOUNDS). The epidemiology cites a study of lead exposed workers which indicated an increased incidence of gastric and lung cancer for battery plant workers only.

None of the MSDSs that were used for the other metals listed a carcinogenicity hazard. There were a lot of data related to hazards due to dusts, powders, and fumes for all the metals. Often only one type of MSDS seemed to be available which covered several forms such as distinct solids (granular, chips, pellets), powder, dust, and fume. A few were more specific such as lead shot, aluminum other than powder, and copper wire. Exposure limits stated in mg/m^3 were found for chromium, copper, lead, nickel, tin, and zinc. Other topics that students were asked to consider related to acute toxicity, skin contact, and water reactivity.

Results of Initial Student LCSSs. The experiment involved weighing a certain amount of the metal, transferring the metal to a graduated cylinder containing some water (no MSDS was supplied for water), noting the increase in volume, and determining the density. A review of the factors that the students hopefully would think about when writing their LCSS include the risk of carcinogenicity, the possibility of toxic or allergic effect from touching the metal, the possibility of reaction of the metal with water or air, and the risk of exposure to dust or fumes of the metal. The safety notes provided with the experiment made no mention of any hazards from the metals other than the usual caution "Wear your goggles at all times while in the laboratory". The majority of students decided that the OSHA PEL would not be exceeded, but it was not clear that they all could explain why. On the other hand, most transferred the MSDS hazard information directly to the laboratory procedure without modification. The preparation of additional CHEM 1000 LCSSs were assigned as part of other Pre-Laboratory exercises, together with additional lecture discussion. By the time three or four of these were completed, the majority of the students began to relate the hazards to the experimental procedures.

Exercise 3: Planning of Experiments. *Prudent Practices* suggests that student laboratory manuals "include questions and assignments that involve the student actively in considering the risks, regulations, and waste disposal costs for alternative approaches to the problem under discussion" (20). The manual being used for CHEM 1000 (21) offered an opportunity to design such an exercise. An experiment on Precipitation Reactions called for the use of silver nitrate, potassium chromate, lead(II) acetate, and barium chloride, together with additional reagents. The students were asked to note the LD_{50} and OHSA PEL from the MSDSs for these compounds, as well as the general toxic effects gathered from information in ANSI Sections 3, 8, and 11, and also any evidence of carcinogenicity. It was pointed out that the experiment would have to be modified because of the carcinogenicity of chromate and potential carcinogenicity of lead(II), and also the disposal costs of the toxic heavy metal compounds were significantly higher than those for metals not in this class. Based on appropriate lecture discussion, it was decided to replace the four compounds mentioned with copper(II) chloride, potassium phosphate, zinc sulfate, and strontium chloride, respectively. Students were asked to prepare CHEM 1000 LCSS for these four compounds.

Summary Discussion. The experiment-specific LCSSs eventually could be prepared to cover all the experiments used. A set of these could be distributed at the beginning of the course, or along with each experiment. These would offer safety information in more detail than that provided by most lab manuals, but not as extensive as, although more useful for the specific experiment than, that from the MSDSs. This approach could be adapted to any chemistry course. If students were introduced to the concept of the LCSS in the elementary course, they could be expected to be familiar with the process for advanced courses. The need for MSDSs in the workplace must be emphasized. The LCSS cannot legally replace an MSDS.

156

Literature Cited.

1. *Prudent Practices in the Laboratory;* National Academy Press: Washington, D.C., 1995.

2. *Safety in Academic Chemistry Laboratories;* Sixth Edition, Committee on Chemical Safety, American Chemical Society: Washington, D.C., 1995.

3. ref. 2. pp 43-44.

4. ref. 1. p 32.

5. http://www.uwyo.edu/A&S/chem/safety1.htm

6. http://www-nanonet.stanford.edu/NanoFab/safety/s6MSDS.html

7. http://siri.org

8. a)Lewis, R. J., Sr. *Sax's Dangerous Properties of Industrial Materials,* 8th Ed. Van Nostrand Reinhold, New York, NY, 1992. b) EPA Toxic Substances Data Sheets. gopher://ecosys.drdr.Virginia.EDU:70/00/library/gen/toxics/ c)Environmental Chemicals Data and Information Network (ECDIN). http://ulisse.ei.jrc.it/cgibin_ecd/inter_query d) Stanford Hazardous Materials Index. gopher://portfolio.stanford.edu:

9. ref. 1. p 235.

10. http://www.hhmi.org/science/labsafe/lcss/listing.htm

11. MDL Information Systems, Inc., 14600 Catalina St., San Leandro, CA, 1984/1994

12. Butcher, S. S.; Pike, R. M.; Mayo, D.W.; Hebert, S. M. *J. Chem. Educ.* **1985,** *62*, A-238.

13. Ni pellet MSDS, Johnson Matthey Co., http://siri.org/msds/h/q307/q103143

14. Ni metal MSDS, Fisher Scientific, http://hazard.com:80/fish/fish/5/1118/.html

15. ref. 1, Sec. 3.C.3.3 pp 43-44 discusses carcinogen classes.

16. Cr metal MSDS, Shieldalloy Corp.,http://siri.org/msds/h/q355/q127147

17. Cr metal MSDS, CERAC Inc., http://haz1.siri.org:80/msds/h/q321/q110225

18. Cr MSDS, Fisher Scientific, http://www.fisher1.com/fb/itv?16..f95.2.msf0002.171

19. Lead Metal Shot MSDS, Acros Organics, http://www.fisher1.com/fb/itv?16..acros. 1.msa0009.263

20. ref. 1. p 3.

21. Hall, J. F. *Introductory Chemistry in the Laboratory, 2nd Ed.;* D.C. Heath: Lexington, MA, 1996; pp 93-101.

22. Fessenden, R. J.; Fessenden, J. S. *Organic Laboratory Techniques;* Brooks/Cole: Pacific Grove, CA, 1993; pp 7, 10-23.

Chapter 14

Chemical Health and Safety Concepts for the Chemistry Curriculum

Javid Mohtasham

Environment, Safety and Hazardous Materials Management Program, Mt. Hood
Community College, Gresham, OR 97030

Chemical safety is a primary focus of professional industrial hygiene.
Most standard chemistry curricula cover background information
necessary for chemical health and safety, but do not specifically cite
hygiene principles. There is little use of chemical health and safety
principles even though they can successfully communicate essential
chemical principles. Safety concepts are easily included as a meaningful
component of chemical education. Incorporation can be achieved by
consistent use of safety concepts coupled with an "introductory" chapter.
This paper outlines topics and provides examples that can be integrated
throughout a text and/or included in a proposed *Chemical Health and
Safety* chapter.

The lack of emphasis on the relationship between basic chemical principles and chemical
health and safety is a serious deficiency in chemical education. Health and safety is a
valid application of chemical science and is a large component of many chemistry-
related professions. Failure to directly address health and safety leads to poor chemical
handling when students enter laboratories and other work environments. Ironically,
safety is a basic concern of nearly every person. This direct relationship provides an
excellent opportunity for chemical educators to apply science to the lives of students,
achieving a high level of "buy-in," yet most instructors do not take advantage.

Incorporating chemical health and safety into chemical practice is not difficult, but
chemical educators and private investigators need successful implementation examples.
Generally speaking, scientists working with chemicals in their research or teaching

laboratories are experts in their field. They know from experience how to handle and be safe around chemicals. Unfortunately, in many cases, this knowledge is not directly transferred to students.

It is common for the general public to wonder about potential chemical exposure and the techniques used to assess potential exposure. A student of chemistry is likely to "buy-in" to their chemical education at a higher level if they see a knowledge gain that is directly transferable to relevant issues. Currently this type of information is considered "Industrial Hygiene" and is not standard information in general chemistry. Chemical safety is commonly the responsibility of Industrial Hygiene courses, yet it can be a valuable tool for chemical educators.

How Should Chemical Health and Safety be Presented?

Presentation of chemical health and safety information should focus on direct applicability, while further validating this field of professional chemistry. Chemical health and safety can be delivered as an appendix, problem sets, anecdotal examples, suggested pre-laboratory assignments, integrated throughout a text, and/or as a stand-alone chapter. Regardless of approach, students should receive a consistent message that health and safety are valuable topics in their chemical education. Proposed ideas do not need to draw attention away from recognized chemical principles. Rather, these ideas can be used to augment the essentials of a good chemical education. Once initiated, the practice could be continued in subsequent courses such as Organic, Inorganic and Quantitative Analysis, increasing the sophistication of information.

Typically the field of industrial hygiene (IH) utilizes a high level of chemistry and chemical principles. Airborne concentrations, chemical states, predicting incompatibility and recognizing chemical compounds are chemistry "basics" used in the IH profession. Additionally, many IH analysis techniques use chromatographic methods commonly taught in general chemistry courses. Concepts from this field are good targets for integration into general chemical education.

General Chemical Health and Safety Topics. Table I and Table II list suggested concepts for college-level General Chemistry semesters I and II, respectively. The issues listed can be integrated throughout any general chemistry textbook, and added to the indices of other chemistry texts. The list does not specifically deliver chemical health and safety principles as "hard science" addenda to each chapter, but serves only to highlight relationships that can be further developed. Specific examples follow later in this chapter.

Stand-alone *Chemical Health and Safety* Chapter. Referring to the subjects in Table I and Table II, a suitable format might be a stand-alone chapter following an introductory chapter. This would directly focus an emphasis on the importance of chemical health and safety to students. Formal chapter delivery would also facilitate the transition to laboratory work, where students are often expected perform with *a priori* safe handling

Table I. Quick Chemical Health and Safety Concepts for Semester I.

1. Chemical Foundations	Dimensional analysis of chemical spill or exposure amount.
	Why oil floats, mercury doesn't: exposures from immiscibles
2. Atoms, Molecules and Ions	Chemical state dictates the hazard (Cl_2, Cl^-, bound Cl).
	Naming compounds: expected components from dissociation.
3. Stoichiometry	% composition of cyanide, arsenic or other.
	Predicting the formation of dangerous products.
4. Rxns and Sol'n. Stoichiometry	Diluting acids or bases to a safe Molarity.
	Predicting which solutions conduct electricity.
5. Gases	Calculating gases in your breathing zone.
	Handling dangerous gases: how T is involved.
6. Thermochemistry	The chemistry of first aid hot and cold packs.
	Predicting heat evolution during spill neutralization.
7. Atomic Structure and Periodicity	Why alkali metals react with halogens.
	What dangers are associated with inert, noble gases.
8. Bonding: General Concepts	Predicting safe dissociation of compounds in water.
	Predicting compound stability from bond type.
9. Covalent Bonding: Orbitals	Why transition metals are reactive: orbital theory predictions.
	Examples of when covalent bonds provide low reactivity.
10. Liquids and Solids	Freeze separation of liquids; sublimation cautions.
	Using vapor pressure to calculate exposures.
11. Properties of Solutions	Gas evolution from liquids (T, P relationships).
	Flash points and boiling points: determining "flammability"
12. Chemical Kinetics	Using catalysts to destroy hazardous wastes.
	Predicting reactions from spontaneity at RTP.
13. Chemical Equilibrium	Equilibrium pressures: storage vessel selection.
	Pressure releases that increase toxic gas evolution.

Chapter outline modeled after Steven Zumdahl's *Chemistry* (1).

Table II. Quick Chemical Health and Safety Concepts for Semester II.

14. Acids and Bases	Corrosives: material and skin effects.
	Measuring neutral solution products.
15. Applications of Aqueous Equilibria	Blood as a buffer system.
	Reactivity of strong and weak acids and bases.
16. Spontaneity, Entropy and Free Energy	Predicting dangerous spontaneous reactions.
	Respecting chemicals: "hidden' energy content.
17. Electrochemistry	Removing Ag from photographic effluent.
	Batteries can be corrosive, even when "dead.'
18. Representative Elements: Groups 1A - 4A	Atomic radius and health effects predictions.
	Welding fumes: trace element route of entry.
19. Representative Elements: Groups 4A - 8A	Phosphorous reactivity based on state.
	Forms of nitrogen and related safety.
20. Transitional Metals and Coord. Chemistry	Oxidation states effects within your body.
	How transition metals are regulated.
21. The Nucleus: A Chemist÷s View	Radiation protective measures and design of PPE.
	Electromagnetic spectrum link to exposure safety.
22. Organic Chemistry	Visualizing common flammable hydrocarbons.
	Naming halohydrocarbons and their uses.
23. Biochemistry	Toxicity in short term metabolism.
	Chronic chemical effects on DNA.
Appendices	Conversions between airborne ppm and mg/m^3.
	Air sampling techniques for chemicals.

Chapter outline modeled after Steven Zumdahl's *Chemistry* (1).

knowledge. The chapter should be direct and clear in its title as well as its contents. A likely title is *Chemical Health and Safety*. The title itself alerts students that they need to be informed of health effects and the safety characteristics of chemicals they might handle.

Additional Descriptive Issues for a Stand-alone Chapter. Formalizing a "*Chemical Health and Safety*' chapter would allow increased room to address the issues in Table III. Each topic can be elaborated with specific examples and problems, validating chemical health and safety as a professional chemical science. A large variety of industrial hygiene resources can be sited from this chapter.

Table III. Descriptive Concepts for a "Chemical Health and Safety' Chapter.

1. Recognizing chemical health and safety issues -toxicology basics.
2. Behaving safely with chemicals - knowing routes of entry.
3. Recommended safe handling practices.
4. Introduction to Material Safety Data Sheets (MSDSs)
5. Use of CAS# in safety resources.
6. Physical and chemical characteristics that define chemical safety.
7. Personal Protective Equipment (PPE) principles, value, selection and limitations.
8. Basic laboratory disposal and clean up of hazardous wastes.
9. Emergency action planning.
10. Criteria for assessing exposure to chemicals.
11. How to find chemical safety information.
12. Problem sets based on Industrial Hygiene and safety principles.

Benefits Achieved by Including Descriptive Issues

Many institutions have developed Chemical Hygiene Plans (CHP) in accordance with OSHA's Lab Standard 29 CFR 1910.1450 (2). However, few institutions successfully link student behavior to the CHP. Common issues and policies covering goggles, fire extinguishers, minor first aid, glassware handling, chemical storage, stockroom access, eyewashes and etc. are often key components to limiting liability, yet they are not included in lectures. These policy issues are based on chemical principles so they are valuable components of chemical education. Leaving these issues to be addressed only by laboratory classes belittles their importance in the field of chemistry. It also skips an opportunity to teach chemistry relevant to everyday behavior.

The advent of interactive and lab-lecture teaching facilities points to a need to bridge lecture and lab. Chemical health and safety provides the necessary educational bridge. Explanations of why certain behaviors are required, from a chemists point of view, provides opportunity to teach chemistry by example. The founding principles of a good chemical education are not questioned, rather they are supported through discussion of descriptive issues (Table III.).

Addressing Themes Common to Education and Health and Safety. The founding reasons for including certain concepts in chemical education are often the founding reasons for chemical health and safety. For example, if the intention is to teach students how to predict chemical reactions, then presentation of stoichiometry and orbitals is enhanced by discussions of reactivity and incompatibility. Students should not be instructed how to "make bombs," but they should be able to predict products and assess the environmental or personal health implications. Students learn to assess chemicals and mixtures, achieving the overall intention. Integration of this sort minimizes future accidents and increases safe handling, while delivering essential chemical education. Hopefully students might then learn to directly apply chemistry to their everyday lives. Another example is an explanation of how personal health effects of chemicals are dependent chemical state. Explaining the difference between halogens in ionic, gaseous, and bound states delivers periodicity, chemical naming and free states while simultaneous informing students that a theoretical, proposed "halogen ban" needs to specifically address the form(s) of halogens in question. A Material Safety Data Sheet (MSDS) provides another successful interface between chemical education and chemical health and safety. Students can be asked to assess the content of a given MSDS throughout the semester in order to understand common use of physical properties, naming terminology, reactivity, waste management, flash points, vapor points, boiling points, density, CAS #, professional references and etc. Many "uneducated" employees know a great deal of chemistry as a result of MSDS use. Acknowledgement and critical review of MSDSs is a direct link between students of chemistry and the "real world."

Increased Knowledge of Exposure and Fire Terminology. Chemical terminology can be upgraded to include the definitions provided in Table IV. This terminology can be used in basic dimensional analysis and explanations of gases and vapors. Traditional definitions of Molecular Weight (MW), Boiling Point (BP), Melting Point (MP), Freezing Point (FP), Solubility (Sol), Flash Point (FP), Vapor Pressure (VP), and Specific Gravity (SG) are commonly used in health and safety, and provide an additional common ground. Professional safety personnel and resources commonly refer to all these concepts, all of which implicate a basic understanding of chemistry.

Using Professional References and the Library in Chemistry. Many chemistry departments are adding written reports to student assessments. Safety and environmental health topics provide an excellent opportunity to bridge chemical knowledge with issues, appropriate to a writing assignment. The materials listed in Table V are valuable references in student writing projects. These texts and URLs are good sources of applied chemistry.

162

Table IV. Health and Safety Terminology

Upper Explosive Limit (UEL): the maximum of concentration (% by volume) of a compound in air which will allow an explosion to occur. Any concentration above this value will not explode because the air is saturated with the compound.

Lower Explosive Limit (LEL): the minimum concentration (% by volume) of a compound in air which will allow an explosion to occur. Anything lower than this value is safe.

Permissible Exposure Limit (PEL): the OSHA regulatory limit for the average concentration of a specific concentration to which an individual can be exposed over and 8-hour period.

Threshold Limit Value (TLV): non-regulatory in the US (ACGIH); the concentration of a chemical in air to which nearly all individuals can be exposed without adverse effects.

Time Weighted Average (TWA): the concentration of a chemical to which an individual is exposed during an 8-hour workday.

Short Term Exposure Limit (STEL): concentration of a chemical that is safe for exposure over a 15-minute period.

Immediately Dangerous to Life or Health concentrations (IDLH): represents the maximum concentration of a chemical from which, in the event of respirator failure, one could escape within 30 minutes without a respirator and without experiencing any escape-impairing or irreversible health effects.

Table V. Examples of Chemical References that Convey Health and Safety.

1. The Chemical Abstract Publication (ACS).
2. Merck Index (Merck).
3. Handbook of Chemistry and Physics (CRC).
4. Prudent Practices in the Laboratory (National Research Council).
5. Safety in Academic Chemistry Laboratories (ACS).
6. Laboratory Health and Safety Handbook (John Wiley & Sons).
7. Safety in the Chemical Laboratory (ACS).
8. Hazardous Laboratory Chemicals Disposal Guide (CRC).
9. Chemistry of Hazardous Materials (Prentice Hall).
10. NIOSH (National Institute for Occupational Safety and Health) Guide Book.
11. ACGIH (American Conference of Governmental Industrial Hygienists) Guide.
12. National Safety Council Publications.
13. Federal OSHA (Occupational Safety and Health Administration) web page.
14. EPA (Environmental Protection Agency) web page.

Sample Problems that Illustrate Chemical Principles

The following eight problems are founded in both chemical health and safety and chemical education. It is recommended that additional examples be developed using

Table I and II as a guideline. Delivery can be in-laboratory assignments, pre-laboratory assignments, example problems within a given chapter, part of a stand-alone chapter, or part of a problem set.

1. The Periodic Table. Students are asked to correlate lead, mercury and selenium occurrence on the Periodic Table with airborne exposure risk. Students are asked to name the chemical, predict the common form at RT, and postulate on a potential periodic trend. The relationship between periodicity and exposure is explored. Exposure limits are found in the NIOSH Guide Book (Table V). Students learn to assess and predict from the periodic table. Future, detailed chemical education on using the Periodic Table is emphasized.

 2. Dimensional Analysis. Students use exposure limits to track units in a conversion that emphasizes dimensional analysis. Students are asked to calculate if an exposure level is exceeded (5 ppm TWA) for measurements of 10 ppm for 3 hours and 3 ppm for 5 hours. Adding the multiplied exposure X hour data and dividing the sum by 8 hours (TWA) requires tracking unit cancellation (time) for a final answer of 5.6 ppm. This is a basic principle drilled early in chemical education.

3. Dilutions and % Composition. Students use an MSDS to record the percent composition (%) of an acid mixture used in metal etching. Students are then asked to calculate how many milliliters of a specific acid (HCl) are contained in the mix given a working volume (V) of 5L. Students must multiply %HCl times V to calculate L, then track conversion to milliliters. This is a common calculation in both lab and lecture, but the use of the MSDS helps legitimize the practice to a novice.

4. Percent vs. Parts Per Million vs. mg/m^3. Students are asked to use a newspaper account of "percent composition in air" (%) for a chemical. Students convert % to parts per million (ppm) and milligrams per cubic meter (mg/m^3) to calculate occupational exposure levels using OSHA's and ACGIH's (Table V) published terminology. Conversion to ppm uses 1% = 10,000 ppm, which can be expanded to track units. Conversion to mg/m^3 requires multiplying the ppm times molecular wieght (MW) and dividing by the STP gas volume 24L (final conversion is mg/m^3 = ppm X MW / 24L). Students can then use the two results to compare the newspaper account to established exposure limits, identifying if acceptable exposure limits have been exceeded. Introduction to gas law is achieved, or this can be used within the gas law unit.

5. Molecular Geometry. Students are asked to draw relevant molecular models for methyl chloride, dichloromethane and hydrazine. The models are then used to visualize the filtration process within a respirator cartridge. Predictions can be made about the relationship between molecular configuration in space and the necessary filtration media. Students can use the NIOSH Pocket Guide to list the specific cartridge types required by OSHA. This introduces the concept of chromatography, linear versus tetrad modeling and handling precautions!

6. Paper Chromatography. This example is based on safety handling of laboratory chemicals, but is useful in lecture. Students are asked to calculate the overall gas composition in a laboratory room where solvent is left uncapped. Students have a room volume to work with, assume full evaporation, and assume no loss of solvent from the room. If 24 students, each given 50.0 ml of acetone, all forget to cap their beakers of solvent, then what is everyone's exposure? This can be as difficult as the instructor likes! Students can be asked to use/calculate evaporation rate, vapor pressure, volatility depending on the instructor's approach. The easiest example is to assume full evaporation within the lab session and complete build-up. The conversion of liquid to gas and the volume occupied allows the student to calculate ppm. The calculated ppm value is then compared to acceptable limits. This exercise can generate a lot of discussion about air quality, be sure to make students use calculations to prove their points!

7. Standardization of Sodium Hydroxide. Students conducting a NaOH titration laboratory are asked to calculate the airborne exposure resulting from an accidental break of 100 ml of a 1M NaOH solution. Assuming rapid formation of vapors, have students calculate the immediate exposure, then compare the results to NIOSH and ACGIH values for an IDLH environment. Ask if students would be hurt by this accident? A healthy discussion of safe handling procedures and emergency action may follow. The use of volatility to determine potential exposure is a focus of this exercise. Students gain a better understanding of "chemical fear" and when it is or is not justified.

8. Hydrofluoric Acid Reactivity with Bone. Students are presented with an exposure story regarding hydrofluoric acid. A description of skin absorption, lack of burning sensation and bone composition (calcium carbonate) is provided. Students are then asked to balance a simple neutralization reaction between the acid and the bone suing proper nomenclature and stoichiometry. Discussion of why hydrofluoric is not rated as corrosive is facilitated, as well as a prediction of what will happen at the tissue level (bone) when this exposure occurs. MSDSs are a good source of reactivity information for this exercise.

Using health and safety within the chemistry curriculum is not hard. There are many examples of applied chemistry related to safety principles. Including these issues in daily behavior, language and teaching sends an important message to students. Scientific integrity is not compromised by this effort, in fact it can be legitimized. Try some of the suggested ideas and see if student commitment to chemistry increases.

Reference

1. *Chemistry*; Zumdahl, Steven S.; D.C. Health and Company; Third Edition; Lexington, MA; 1993.

Chapter 15

Testing for Chemical Health and Safety Understanding

Lucy Pryde Eubanks

ACS DivCHED Examinations Institute and Department of Chemistry, Clemson University, Clemson, SC 29634–1913

There is one sure way to help guarantee that chemical health and safety concerns are taken seriously - test for them! Every student has at some point asked the question–"Is it going to be on the test?"– when what was really being asked was whether or not the instructor or supervisor valued the importance of the information. Although tests are sometimes criticized for driving curricula, in fact this can be a positive benefit in shaping an increased awareness of the importance of chemical health and safety. This chapter will offer guidelines for designing assessment that can be used in the classroom to help measure student learning not only about safety, but also about concepts of chemistry. The importance of setting the goals, criteria, purposes, and types of assessment will be explained, with examples given to illustrate their applicability to chemical health and safety. Testing may also take the form of a standardized examination, such as that prepared by the ACS DivCHED Examinations Institute.

The rapid growth in academic safety programs has been driven by need as well as mandated by external regulation. No matter how students use their chemistry backgrounds in the future, safety will be expected to be part of their performance and attitude. Just as there are many reasons for the increased emphasis on academic programs, there are many approaches to achieving a "safety culture" within academic chemistry departments. Industrial employers want to save much of the training time required to bring new employees into an acceptable level of working safety knowledge. OSHA lab standards must be understood by graduates if they are to be professional chemists. Teaching and research faculty must share the responsibility to incorporate safety education into their curricula. Even if one makes the assumptions that all faculty already understand the need to educate for the OSHA-savvy chemist, have the necessary knowledge, and will incorporate chemical health and safety into their curricula, this question remains: how will you *know* that students have learned what you feel *they should know* about chemical health and safety? This is a most important question. It is not enough to simply *expose* students to ideas about chemical health and safety. Faculty must systematically incorporate chemical health and safety throughout the curriculum, not just once at the start of the first semester. Faculty must actively investigate how effectively students *learn* concepts, skills and attitudes in this relatively new curricular area. Finally, the faculty must also assume the responsibility to communicate their research-based findings on safety.

Testing for Chemical Health and Safety Knowledge

There is one sure way to help guarantee that chemical health and safety concerns are taken seriously by students at all academic levels, and that is to test for these concerns! Every student has at some point asked the question–"Is it going to be on the test?"–when what was really being asked was whether or not the instructor or supervisor *valued* the importance of the information, concept, or skill being discussed, demonstrated, or modeled. Although tests are sometimes criticized for driving curricula, in fact tests can be a positive benefit in shaping an increased awareness of the importance of chemical health and safety. Most academic chemists are not experts in the processes of evaluating student learning in the area of chemical health and safety, and it is by no means a simple matter. Certainly students are expected to learn specific information and understand the reasons behind many rules and regulations, but there is much more to the issue. The instructor also expects students to change attitudes towards their personal safety in the laboratory, and to understand the need for an entire team to "buy in" to the safety culture. How is it going to be possible to find out if students have in fact changed in all of these sophisticated ways? Table I offers some "lessons learned" in comparing previous practice with future assessment practice.

Considerations for Designing Classroom Assessments

Fortunately, many of the same questions asked in designing all chemistry evaluation instruments are applicable to the safety area of academic chemistry programs as well. What is it that academic chemists ideally think about before designing any test for use with their students? What are some of the assessment issues to consider when deciding how to carry out such testing?

Goals. While many factors could be considered, the most important question to ask is what specific **goals** were set in teaching about chemical health and safety? Careful consideration of this question will not only focus teaching but will also make designing assessment instruments far easier. All instructors are probably much more comfortable with setting goals within a traditional chemistry sub-discipline specialty than will be the case within the wider range of chemical health and safety. Why is it that there is confidence when asking questions about pH or solubility or electrochemistry or nuclear decay equations, but that it initially seems more difficult when the goal is "to educate students about safely using and storing organic chemicals?" Part of the uncertainty may be unfamiliarity with the particular safety concept, but this uncertainty can be eased by narrowing the goals so they are measurable, not global.

Guidelines of the ACS Committee on Professional Training state that: "Discussion of current health and safety issues must be an integral part of the chemistry curriculum,

Table I. Lessons Learned–Past and Future Assessment Practices

PREVIOUS PRACTICE	FUTURE PRACTICE
• Assuming there is transference from large wall charts into prepared minds	• Carrying out assessment of chemical health and safety knowledge, skills and attitudes on a regular basis
• Assuming safety information should be delivered only in the laboratory	• Integrating chemical health and safety information in all aspects of a course
• Assuming only experts can write appropriate chemical health and safety assessment items	• Taking responsibility for continued student assessment as part of educating the OSHA-savvy chemist
• Assuming identification of hazards and risks will lead to appropriate strategies for minimization and avoidance	• Discussing the hazards and risks as well as benefits and burdens, leading to the ability to make informed decisions
• Assuming students need only minimal safety instruction and they retain disconnected individual factoids	• Assessing what is most highly valued, such as critical thinking skills needed to place factoids in context

beginning early in the core courses…". *(1)* To help achieve this, they suggest that the discussions include topics such as those given in Table II.

Table II. Topic Areas for Inclusion in Undergraduate Chemistry Courses

• Acute and chronic toxic effects of chemicals

• Flammability of chemicals

• Explosive character of chemicals

• Radiation hazards of chemicals (if applicable)

• Pertinent OSHA, state and local regulations

• Recognition of hazards and actions required for possible mitigation or avoidance

168

Many of these areas are too broadly defined to be immediately useful as goals for assessment items in a given course, but they do indicate the general areas of interest. To make these goals *measurable*, they must be considered by each teacher in light of the particular curriculum being followed. Table III illustrates the difference between broadly defined goals such as those set out by CPT and goals with a more narrow focus, making them more immediately useful in applicable situations.

Table III. Sample Goals for Assessment of Chemical Health and Safety Topics

GLOBAL GOALS	FOCUSED GOALS
• A student should have knowledge of the physical and chemical properties of chemicals handled.	• A student should know why 1-propanol can be substituted for chloroform and/or dichloromethane in the extraction of caffeine from tea.*(2)*
• A student should know how to comply with OSHA regulations concerning the safe use of chemicals.	• A student should be able to demonstrate understanding of the information in a material safety data sheet (MSDS) for 1-propanol.
• A student should know the flammability of chemicals.	• A student should be able to explain under what conditions 1-propanol is flammable in air.
• A student should know the location of all safety equipment.	• A student should be able to locate safety showers and fire extinguishers and communicate under what circumstances each item would be used in a laboratory fire.

Criteria. The next question to consider is what **criteria** are appropriate to enable the teacher to determine the degree of student success? Is complete mastery of a particular topic desired, or is the expectation that students will show incremental gains appropriate to their age and educational experience? Will students be expected to demonstrate their abilities to follow an algorithm or to process information conceptually? Will credit be given only for the "right answer" or will the student be given the opportunity to display each step of a critical thinking process?

To better understand what is meant by choosing appropriate criteria for assessment, consider one of the most common pieces of safety equipment, the safety hood. What is it that students are expected to know and understand about the proper use of a hood, no matter where it is found? Table IV brings the issue of changing criteria into better focus.

Table IV. Contrasting Questions With Different Criteria for Success

ONE CRITERION FOR SUCCESS	MULTIPLE CRITERIA FOR SUCCESS
How can you tell if your hood is working properly? (A) By performing a smoke test (B) By keeping the sash at a preset position (C) By checking a properly functioning airflow monitor (D) By checking the test label	What are the guidelines for proper use of the chemical hood in your laboratory?
Comments: In this single-answer, multiple-choice format, there is just one right answer, making the question easy to grade and to produce statistics for a large population group. Most students in introductory chemistry either know that the hood must be checked with an airflow monitor, or can easily guess from the way in which the question is worded. (The correct answer is the longest and contains qualifying modifiers.) In fact, this question is likely to be answered correctly even by individuals with no knowledge of how the performance of hoods is determined. This may even have been the purpose of this question, as it came from a *Chemical Health & Safety* Reader Survey designed as a contest with the prize being a free ACS safety video. *(3)*	**Comments:** This question requires that a student think about and then explain as many of the parameters about the use of a chemical hood as are appropriate to their present experience. A recent article in *Chemical Health & Safety* listed 11 different guidelines dealing with certification, maintenance, and safe operation *(4)* and further suggestions can be found in other ACS publications *(5)*. It would even be possible to turn this open-ended question into a *multiple*-answer, multiple-choice, machine scorable question by listing the possible true or false statements the student might make. Then the question would ask students to choose which statements are appropriate guidelines and which are not.

Purpose. Another important question to ask is what is the **purpose** of the evaluation? Will the results be used to make changes in teaching practices? Will the results be used to make judgments about the merits of a particular curricular approach to chemical health and safety? Will the results be used to make judgments about the relative merits of student learning within a class, or to compare students to a nationally determined norm? Some of the considerations about the purposes of assessment are given in Table V.

Table V. Characteristics of Tests Designed for the Different Purposes of Measurement and Instruction[a]

ASSESSMENT FOR MEASUREMENT	ASSESSMENT FOR INSTRUCTION
• Valid, Reliable	• Quality judged by effect on instruction
• Objective, Efficient	• Design determined by instructional purpose intended
• Centrally mandated	• Teacher mandated
• Widely applicable	• Applicable to local context
• Multiple Choice (Single answer)	• Format determined by instructional value
• Machine-scored centrally	• Locally scored
• Delayed feedback	• Immediate feedback
• Used independently	• Used with other information
• Stable scores	• Scores affected by short-term learning
• Results designed for external user	• Results meaningful to teachers and students

[a] SOURCE: Adapted from reference (6), p. 100. For those wanting further elaboration about the differing purposes of assessment, consult references (6–8).

For the education of the OSHA-savvy chemist, the primary purpose for assessment will usually be instruction. The teacher will be able to set assessment tasks and evaluate their results on the local level. However, there certainly will be times when it is useful to take some external measure of student success with learning the many aspects of the safety culture. External measures may take the form of a paper-and-pencil examination, such as the one described later in this chapter that has been produced by ACS DivCHED Examinations Institute. The external measure may also take the form of a performance evaluation administered by a supervisor, and such evaluations may be used in either academia or industry to satisfy a particular gate or barrier to professional advancement.

One of the most difficult purposes associated with evaluating chemical health and safety knowledge is to use the assessment instrument for some type of certification. Certification exams for Chemical Hygiene Officers, such as the exam being prepared through the National Registry in Clinical Chemistry (9), have legal ramifications that must be carefully considered throughout their preparation, standardization, and administration. Think, for example, of designing a chemical health and safety certification exam...but one that is seriously flawed, making the interpretation of the results unclear. There also are issues of security that are important in most high-stakes examinations.

Type. What **type** of assessment will be used? Many instructors are willing to consider an ever-widening array of question types, delivery systems, and scoring methods. Will the student be asked to complete a traditional paper-and-pencil test? Will only single - answer multiple-choice questions be used, or will multiple-answer questions be possible? What about alternatively designed questions, such as grid or linked questions? Will open-ended essay questions be used? Will the questions asked by algorithmic in nature or more conceptual? Table VI compares these last two question types.

Table VI. Comparison of Question Types: Algorithmic and Conceptual

ALGORITHMIC QUESTION	*CONCEPTUAL QUESTION*
What does this symbol mean? **(A)** corrosive **(B)** explosive **(C)** flammable **(D)** oxidizer	An iron garden tool and gasoline both can undergo oxidation in the atmosphere. There are not many regulations about storing an iron garden tool, but there are for storing and transporting gasoline. Discuss the properties of each material to explain why there is this regulatory difference.
Comments: Most students either know that this symbol represents flammable materials, or can easily guess from the nature of the symbol. The answers may be given in a multiple-choice format, or left open-ended for the student.	**Comments:** This question requires that a student think about and then explain the differences in rapid and slow oxidation and relate them to properties of toxicity, flammability, and explosiveness of gasoline. It would even be possible to turn this into a multiple-answer machine scorable linked question by listing the possible true or false statements the student might make and asking the students to choose which statements support the regulatory difference.

Another type of question that is used with increasing frequency in modern curricula is one that uses information represented in graphs or grids as the basis of decision-making on the part of the students. Particularly within the area of chemical health and safety, there is a large variety of data that can be represented in this manner in the scientific and regulatory literature, as well as in the press and on the news. Students seem particularly interested in data if it represents relatively current information. If data are used for assessment purposes, they should usually represent new information, but related to topics that they are prepared to process *(10)*. Table VII gives an example of current data displayed in graphical format that can be used for a wide variety of assessments.

Table VII. Questions Based on Graphical Representations *(10)*

Use of Toxic Chemicals	Production of Toxic By-Products

Sample Question:

Based on the two graphs in Table VII, one can conclude that

(A) between 1990 and 1995 there was a 20% reduction in both the use of toxic chemicals and in the production of toxic by-products.

(B) in 1993 the ratio of toxic by-products produced to toxic chemicals used was the lowest of the years shown.

(C) the ratio of toxic by-products produced to toxic chemicals used improved from 0.12 in 1990 to 0.11 in 1995.

(D) between 1990 and 1993 the ratio of toxic by-products to toxic chemicals used decreased; the ratio increased after 1993.

Comments: This single-answer multiple-choice question requires more than simply reading the graphs. Processing must be done to compare the information, leading to a question that is statistically able to discriminate among student responses. Choices must be carefully crafted so that plausible, but incorrect, conclusions might be chosen by some of the students. (**C** is correct)

Sample Question:

What conclusions can be drawn from the comparison of these two graphs?

Comments: This question requires that a student examine the graphs carefully and process the information to draw conclusions about the link between the use of toxics and toxic by-products. By leaving it open-ended, there is the possibility of a large variety of responses; scoring can be eased with the development of a scoring rubric. This open-ended question could also be a group activity. The *criteria* for scoring responses can vary dramatically. For example, students might well note that while there is a general trend linking toxic chemical use with toxic waste in the state, one cannot account for all of the reduction based on decreased usage. If this question were being asked after students had studied any of the process changes associated with "green chemistry", information about the chemistry of the substitutes could be expected. Incorrect conclusions should be avoided too, such as stating that results can generalized out from just this information to the national level.

As can be seen from these examples, there are many possible decisions that faculty must make in choosing appropriate classroom assessment. In addition to the ones already discussed, there are decisions about whether the students will work alone or in cooperative learning groups. Access to calculators and reference material during the assessment must be considered. Will students be able to use the same materials consulted by professionals, such as Material Safety Data Sheets or reference books? Can the evaluation be carried out with the use of computers, or from the Web? Will the student be asked to actually demonstrate safe procedures in the laboratory, carrying out a performance-based assessment? Will the student be given the opportunity to display creativity by carrying out evaluation in less traditional settings, perhaps by designing a "safety skit" and performing it for the class? All of these options can be considered in developing an assessment repertoire suitable for classroom assessment. Table VIII provides a summary of previous and future practices in design considerations when starting to evaluate chemical health and safety knowledge and skills.

Table VIII. Lessons Learned–Designing Assessment By Considering Goals, Criteria, Purpose and Type

PREVIOUS PRACTICE	*FUTURE PRACTICE*
• Neglecting to think carefully about the **goals** of any evaluation being designed	• Considering **goals** appropriate to your chemical health and safety curriculum
• Choosing inappropriate **criteria** by which to evaluate the students being tested	• Choosing **criteria** that enable the teacher to understand the student's reasoning processes
• Failing to define and communicate the **purpose** of the evaluation	• Defining and communicating the **purpose** or purposes of the evaluation
• Making repetitive or limiting choices about the **type** of assessment used	• Expanding the repertoire of **types** of assessment experiences for the student

ACS DivCHED Examinations Institute Standardized Examination

Integrating chemical health and safety into the undergraduate curriculum can be accomplished through classroom testing, but a complementary approach is to develop external standardized testing instruments. A standardized test allows impartial comparison of the students or technicians to a national norm and removes the burden from the shoulders of the local teacher or supervisor of preparing such a test. It also provides a vehicle for communication so that the goal of increased knowledge about chemical health and safety concerns can be achieved. The purposes of developing an

ACS examination are often related to the idea of validating the importance of a new or changed curriculum, as well as responding to the needs of the chemical education community. It became painfully apparent to Institute personnel in the production of the *ACS Test-Item Bank for General Chemistry (12)* that during the previous 60-year history of producing general chemistry test questions, almost none dealt with issues of chemical health and safety. In the present collection of nearly 1400 questions, there are only three that deal with any aspect of safety; two of those are related to the proper methods for diluting concentrated sulfuric acid! The statistics are similar for the *ACS Text Item Bank for High School Chemistry (13)*.

In the fall of 1994, a committee of the ACS DivCHED Examinations Institute started to develop a national standardized examination in chemical health and safety. Acknowledging that no one test would be suitable for all students, the committee decided to make their first test for advanced students such as those serving as teaching or research assistants, as well as for technicians working in industrial or government laboratories. As with classroom assessment, the next step was to consider the goals of the assessment instrument. What should an advanced student or practicing technician be able to demonstrate about their knowledge in the area of chemical health and safety? Table IX gives a list of assessment goals developed by the committee.

Table IX. Assessment Goals for the ACS Chemical Health and Safety Examination *(14)*

A student should be able to:

- demonstrate knowledge of safety vocabulary
- know properties and interactions of hazardous materials
- explain the reasons for safety procedures
- select appropriate equipment for a specific task
- explain the functions of varied safety equipment
- understand the basis of chemical health and safety hazards
- know the meaning of safety symbols
- understand labeling systems
- prioritize emergency actions
- interpret safety–related acronyms
- understand safe storage and disposal procedures
- demonstrate knowledge of regulatory processes
- communicate the characteristics of a safe work place
- relate the characteristics of a chemical hygiene plan

As can been seen from this list, some of these goals involve simple memorization or the use of simple algorithms, but others are far more complex and not easily measured. While the committee felt it was important to identify and reward significant knowledge and critical thinking skills, they also felt constrained by the need to produce a machine-scorable examination that would be easy to distribute, score, and use to calculate national norms. The committee soon found that while it was easy enough to ask an algorithmic question, it was much more difficult to frame questions requiring complex reasoning, although this is certainly not impossible within a multiple-choice format. These difficulties are not surprising, for committees working throughout the 70-year history of the Exams Institute program have found exactly the same thing to be true! Although it took nearly three years of discussion, editing, reworking, and more editing, the committee successfully completed their work in the fall of 1996, and this 75-question single-answer multiple-choice examination is now available from the Exams Institute. *(15)* National norms will be calculated as sufficient data has been returned to the Institute. There are preliminary plans to develop examinations for other population groups such as introductory chemistry students at both the high school and college levels.

The Challenge of Change

The use of chemical health and safety education as an effective teaching tool within the chemistry curriculum represents an important change in thinking for most educators. The thesis of this volume is that chemical health and safety considerations can be fully integrated into chemistry courses and used as a potent way to interest students, to teach chemical concepts, and to achieve the overall goal of all education – to improve student learning. This chapter has offered some suggestions for both internal and external assessment, all targeted to help determine if innovations in the chemical health and safety curriculum are matched by actual gains in student knowledge and attitudes. Table X focuses attention on previous unsuccessful practices that must be considered targets for immediate change and contrasts those with future practices likely to enrich learning of chemical health and safety ideas and attitudes.

As more instructors become active participants not only in teaching about chemical health and safety, but measuring student learning in chemical health and safety, it is important to remember that effecting change in assessment is not the easiest of tasks. The lowest energy pathway is clearly just to test as we have been tested. Change requires investment of time and resources, models for guidance, personal effort, and support from one's peers. However, change is well worth the effort, for it will be possible to gain more information about students learning and about teaching practices. Everyone involved in the teaching and learning process will gain from promoting chemical health and safety knowledge, skills and attitudes.

Table X. Lessons Learned–Effecting Change in Assessment of Chemical Health and Safety Knowledge

PREVIOUS PRACTICE	FUTURE PRACTICE
• Assuming one exposure works for instilling chemical health and safety knowledge and attitudes	• Adopting a continuous practice of both instruction and assessment
• Separating the instruction and the evaluation from the real world of chemistry	• Utilizing examples from academic, industrial, and consumer settings
• Neglecting to convey the paths to resources on chemical health and safety	• Requiring information from the Web and print resources to be used during evaluations
• Emphasizing individual assessments prepared by external examiners	• Encouraging students to take a more active role in their learning and self-evaluation
• Ignoring research on teaching and learning, including learning styles	• Providing all students with the opportunity to demonstrate success

References

1. *Undergraduate Professional Education in Chemistry: Guidelines and Evaluation Procedures;* ACS Committee on Professional Training; American Chemical Society, Washington, DC, 1992; p 10.
2. Murray, S.D.; Hansen, P.J.; *J. Chem. Educ.* **1995**, 72, p 851.
3. *Chem.Health.Saf.* **1996**, 3(2), p 32.
4. Gershey, E. L., et al; *Chem.Health.Saf.* **1996**, 3(6), p 38.
5. ACS Committee on Chemical Safety; *Safety in Academic Chemistry Laboratories, 6th Ed;* American Chemical Society, Washington, DC, **1995**; pp 48-49.
6. Kulm, G.; Malcom, S. M., Eds.; *Science Assessment in the Service of Reform*; American Association for the Advancement of Science: Washington, DC, **1991**.
7. Angelo, T. A.; Cross, K. P.; *Classroom Assessment Techniques: A Handbook for College Teachers, 2nd Ed.*; Jossey-Bass Publishers: San Francisco, CA; **1993**.
8. Eubanks, I. D.; Eubanks, L. T.; *Writing Tests and Interpreting Test Statistics: A Practical Guide*; ACS DivCHED Examinations Institute, Clemson, SC, **1995**.
9. National Registry in Clinical Chemistry, 815 15th St., NW, Suite 630, Washington, DC 20005.

10. Chem.Eng.News, June 30, **1997**, p. 27.
11. Eubanks, L. P.; "Assessment of Decision-Making Skills": *Chemistry: The Key to the Future; Proceedings of the Thirteenth International Conference on Chemical Education*, San Juan, PR., **1995**.
12. Eubanks, I. D.; Eubanks, L. T.; *ACS Test-Item Bank for General Chemistry*; ACS DivCHED Examinations Institute, Clemson, SC, **1992**.
13. Eubanks, I. D.; Eubanks, L. T.; *ACS Test-Item Bank for High School Chemistry*; ACS DivCHED Examinations Institute, Clemson, SC, **1993**.
14. Eubanks, L. P., *Chem.Health.Saf.* **1996**, *3,* pp. 10-12.
15. *Chemical Health and Safety Examination, Form 1997.* ACS DivCHED Examinations Institute; Clemson University, Clemson, SC, **1997**.

INDEXES

Author Index

Subject Index

Bestsellers from ACS Books

The ACS Style Guide: A Manual for Authors and Editors (2nd Edition)
Edited by Janet S. Dodd
470 pp; clothbound ISBN 0–8412–3461–2; paperback ISBN 0–8412–3462–0

Writing the Laboratory Notebook
By Howard M. Kanare
145 pp; clothbound ISBN 0–8412–0906–5; paperback ISBN 0–8412–0933–2

Career Transitions for Chemists
By Dorothy P. Rodmann, Donald D. Bly, Frederick H. Owens, and Anne-Claire Anderson
240 pp; clothbound ISBN 0–8412–3052–8; paperback ISBN 0–8412–3038–2

Chemical Activities (student and teacher editions)
By Christie L. Borgford and Lee R. Summerlin
330 pp; spiralbound ISBN 0–8412–1417–4; teacher edition, ISBN 0–8412–1416–6

Chemical Demonstrations: A Sourcebook for Teachers, Volumes 1 and 2, Second Edition
Volume 1 by Lee R. Summerlin and James L. Ealy, Jr.
198 pp; spiralbound ISBN 0–8412–1481–6
Volume 2 by Lee R. Summerlin, Christie L. Borgford, and Julie B. Ealy
234 pp; spiralbound ISBN 0–8412–1535–9

The Internet: A Guide for Chemists
Edited by Steven M. Bachrach
360 pp; clothbound ISBN 0–8412–3223–7; paperback ISBN 0–8412–3224–5

Laboratory Waste Management: A Guidebook
ACS Task Force on Laboratory Waste Management
250 pp; clothbound ISBN 0–8412–2735–7; paperback ISBN 0–8412–2849–3

Reagent Chemicals, Eighth Edition
700 pp; clothbound ISBN 0–8412–2502–8

Good Laboratory Practice Standards: Applications for Field and Laboratory Studies
Edited by Willa Y. Garner, Maureen S. Barge, and James P. Ussary
571 pp; clothbound ISBN 0–8412–2192–8

For further information contact:
Order Department
Oxford University Press
2001 Evans Road
Cary, NC 27513
Phone: 1-800-445-9714 or 919-677-0977
Fax: 919-677-1303

Highlights from ACS Books

Desk Reference of Functional Polymers: Syntheses and Applications
Reza Arshady, Editor
832 pages, clothbound, ISBN 0–8412–3469–8

Chemical Engineering for Chemists
Richard G. Griskey
352 pages, clothbound, ISBN 0–8412–2215–0

Controlled Drug Delivery: Challenges and Strategies
Kinam Park, Editor
720 pages, clothbound, ISBN 0–8412–3470–1

Chemistry Today and Tomorrow: The Central, Useful, and Creative Science
Ronald Breslow
144 pages, paperbound, ISBN 0–8412–3460–4

Eilhard Mitscherlich: Prince of Prussian Chemistry
Hans-Werner Schutt
Co-published with the Chemical Heritage Foundation
256 pages, clothbound, ISBN 0–8412–3345–4

Chiral Separations: Applications and Technology
Satinder Ahuja, Editor
368 pages, clothbound, ISBN 0–8412–3407–8

Molecular Diversity and Combinatorial Chemistry: Libraries and Drug Discovery
Irwin M. Chaiken and Kim D. Janda, Editors
336 pages, clothbound, ISBN 0–8412–3450–7

A Lifetime of Synergy with Theory and Experiment
Andrew Streitwieser, Jr.
320 pages, clothbound, ISBN 0–8412–1836–6

Chemical Research Faculties, An International Directory
1,300 pages, clothbound, ISBN 0–8412–3301–2

For further information contact:
Order Department
Oxford University Press
2001 Evans Road
Cary, NC 27513
Phone: 1-800-445-9714 or 919-677-0977
Fax: 919-677-1303